CATALOGUE

1887.

MILLERS FALLS COMPANY

H. L. PRATT, Pres.

L. J. CUNN, Treas,

HARDWARE MANUFACTURERS,

WAREHOUSE,

93 READE STREET,

NEW YORK.

FACTORY,

MILLERS FALLS,

MASS.

THE ASTRAGAL PRESS
Mendham, New Jersey

Library of Congress Catalogue Card Number:
91-78027
International Standard Book Number:
1-879335-27-1

Published by
The Astragal Press
P.O. Box 239
Mendham, New Jersey 07945-0239

"A WORKMAN IS KNOWN BY HIS TOOLS."

The following named tools are such as any workman might be proud of. Our Bit Braces have long been known as the best in use and are now sold in all parts of the world. We also command the Market on Breast and Hand Drills. Our Millers Falls Boring Machine is a new thing, but is taking the entire demand wherever it has been introduced, it is so manifestly better than any other kind. The same may be said of the Auger Handle. Nothing could be more desirable. Everything in this list is fully warranted, and in use; if any part breaks, duplicates can be had by sending to us, as all parts are interchangeable.

We are sending out this Catalogue at a cost of $1,200, because we think the Carpenters ought to know where good tools can be had, when so many poor ones are flooding the market.

If you do not want tools now, please preserve the list and send for them when in need. Hardware Dealers generally keep our goods, and most of those who do not will furnish them if so requested.

In places where these goods cannot be had from dealers, those wanting them may send to us, and on receipt of the price we will send by mail or express, prepaid, any of the goods named on pages 2, 3, 4, 5, 6, 7, 8, 9, 10, 11, 12, 14, 15, 16, 17, 18, 28. Goods named on the other pages will be sent at the cost of purchaser for freight.

If Jack-screws or Vises are wanted, state what kind and how many, and we will name a special price which will be quite a large discount from the prices named in this list.

We have put our Bracket Saws in this Catalogue for the benefit of the boys. Nothing else will educate them so fast in mechanical science and practice as the use of these saws. It is an education of itself, and no urging is needed to keep them up to their work. We shall be glad to hear from all who receive this book.

MILLERS FALLS CO.,

74 CHAMBERS ST.,

NEW YORK.

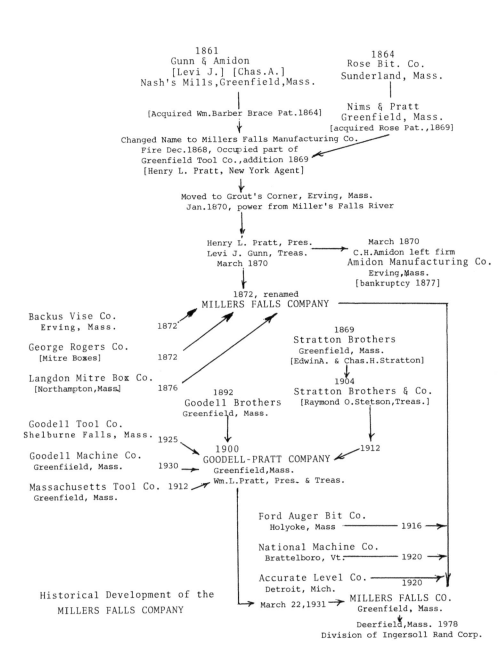

1861
Gunn & Amidon
[Levi J.] [Chas.A.]
Nash's Mills,Greenfield,Mass.

1864
Rose Bit. Co.
Sunderland, Mass.

Nims & Pratt
Greenfield, Mass.
[acquired Rose Pat.,1869]

[Acquired Wm.Barber Brace Pat.1864]

Changed Name to Millers Falls Manufacturing Co.
Fire Dec.1868, Occupied part of
Greenfield Tool Co.,addition 1869
[Henry L. Pratt, New York Agent]

Moved to Grout's Corner, Erving, Mass.
Jan.1870, power from Miller's Falls River

Henry L. Pratt, Pres.
Levi J. Gunn, Treas.
March 1870

March 1870
C.H.Amidon left firm
Amidon Manufacturing Co.
Erving,Mass.
[bankruptcy 1877]

1872, renamed
MILLERS FALLS COMPANY

Backus Vise Co.
Erving, Mass.   1872

George Rogers Co.
[Mitre Boxes]   1872

Langdon Mitre Box Co.
[Northampton,Mass.]   1876

1869
Stratton Brothers
Greenfield, Mass.
[EdwinA. & Chas.H.Stratton]

1904
Stratton Brothers & Co.
[Raymond O.Stetson,Treas.]

1892
Goodell Brothers
Greenfield, Mass.

Goodell Tool Co.
Shelburne Falls, Mass. 1925

Goodell Machine Co.
Greenfiield, Mass.   1930

Massachusetts Tool Co. 1912
Greenfield, Mass.

1900
GOODELL-PRATT COMPANY
Greenfield,Mass.
Wm.L.Pratt, Pres. & Treas.

1912

Ford Auger Bit Co.
Holyoke, Mass ——————— 1916 ——→

National Machine Co.
Brattelboro, Vt.——————— 1920 ——→

Accurate Level Co. ———————
Detroit, Mich.   1920

Historical Development of the
MILLERS FALLS COMPANY

March 22,1931 ——→

MILLERS FALLS CO.
Greenfield, Mass.

Deerfield,Mass. 1978
Division of Ingersoll Rand Corp.

MILLERS FALLS CO.

Millers Falls Co. was established March 1872 at Erving, Massachusetts, from the earlier firm of Millers Falls Manufacturing Co. Levi J. Gunn and Charles A. Amidon associated under the firm name of Gunn & Amidon at Nash's Mills, Greenfield, MA, in 1861. Formerly both had worked at the Greenfield Tool Co. Gunn & Amidon manufactured a clothes wringer. In 1864 they acquired the Wm. Barber patent and began making braces of this design. Late in 1868 they decided to erect a factory at Grout's Corner, Erving, making use of water power from a dam on Millers River. Their firm name was then changed to Millers Falls Manufacturing Co. Henry L. Pratt was engaged as their New York agent at 87 Beekman Street, later moving to Chambers Street.

A fire in December 1868 destroyed their factory at Greenfield. They continued operations, renting a portion of the Greenfield Tool Co. space. During January 1870 they removed to the new factory at Grout's Corner. In March 1870 Henry Pratt became President and Levi Gunn, Treasurer. Soon after C.H. Amidon left to form his own manufacturing firm. This was also at Millers Falls, but it went bankrupt in January 1877.

In 1872 the firm was renamed Millers Falls Company and the nearby Backus Vise Co. was acquired, as was the George Rogers Co., makers of mitre boxes. In 1876 the Langdon Mitre Box Co. was also acquired.

The 1887 catalog noted that the company's New York sales office had moved again from 74 Chambers Street to 93 Reade Street. The catalog featured Millers Falls' own tools but also carried other tools on consignment: Alford hand vises, Lewis patent single twist spur bits, Johnson's automatic boring tools, and Stratton's patent levels. The Company continued to acquire other firms, expanding its lines of tools.

On March 22, 1931, the firm acquired Goodell Pratt Co., which had previously taken in Stratton Brothers & Co. The headquarters of Millers Falls Co. then returned to Greenfield, occupying the former offices of Goodell Pratt. In 1978 the firm relocated in Deerfield, MA, abandoning the manufacturing sites at Millers Falls and Greenfield. The Company is now a subsidiary of Ingersoll Rand, Inc., and continues to remain among the outstanding hand tool manufacturers.

Kenneth D. Roberts
October 1981

BARBER'S IMPROVED BIT BRACE FORGED STEEL JAWS.

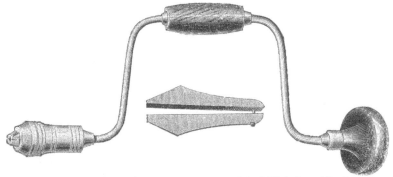

Rolled Steel Sweep, Highly Polished and Nickel-Plated, with Rose-wood Handle and Lignumvitæ Head, with Brass Anti–Friction Collar.

PRICES.   PER DOZ.

No. 10, 14 inch sweep...$33 00
" 11, 12   "   " 30 00
" 12, 10   "   " 27 00
" 13, 8   "   " 24 00
" 14, 6   "   " 21 00
" 15, 4   "   " 20 00
" 16, 4   "   "   { With Chuck, as on our Patent adjustable Tool } 20 00
                          { Holder, for Piano makers' use. }

This is the Barber Brace, with Amidon's Patent Jaws, being the regular goods which we have sold for the last ten years.

BARBER'S IMPROVED RATCHET BRACE.

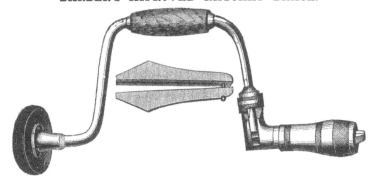

Steel Brace, Nickel-Plated, with Brass Anti-Friction Collar.

This is the perfection of a Bit Brace, having a regular Barber Chuck, with forged steel Jaws, and also Dolan's Patent Ratchet Attachment, to be used in places where there is not room to revolve the sweep. A slight back and forth motion will drive the bit in or out. All the working parts are made of steel, and the whole brace is beautifully finished.

PRICES.   PER DOZ.

No. 30, 14 inch sweep...$42 00
" 31, 12   "   " 39 00
" 32, 10   "   " 36 00
" 33, 8   "   " 33 00

BARBER BRACE.

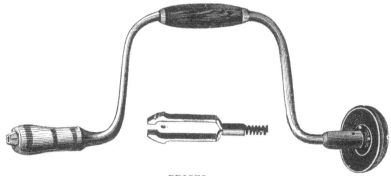

PRICES,   PER DOZ.

No. 40, 14 inch sweep.......................................$33 00
" 41, 12 " " 30 00
" 42, 10 " " 27 00
" 43, 8 " " 24 00
" 44, 6 " " 21 00

The outside appearance of this Brace is exactly the same as the Barber Improved Brace, on page 1, but the jaws are as seen in this cut. The jaws are forged steel, the sweep of rolled steel, heavily nickel-plated. The head lignumvitæ, and the handle rose-wood. Brass Anti-Friction Collar.

BARBER RATCHET BRACE.

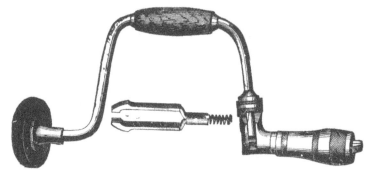

PRICES.   PER DOZ.

No. 61, 12 inch sweep $39 00
" 62, 10 " " 36 00
" 63, 8 " " 33 00

This Brace is exactly like the Ratchet Brace on page 1, except the jaws, which are as seen in this cut. The sweep and jaws are steel, the nickel-plating and trimmings all first-class. Brass Anti-Friction Collar.

DRILL BRACE,

ANTI-FRICTION COLLAR.

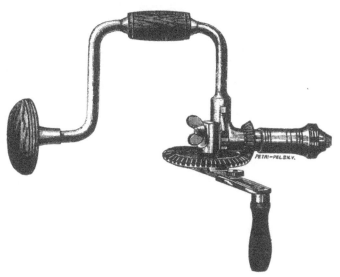

This is a ten-inch sweep Ratchet Brace, with Gear Wheels speeded about three to one, to be used for drilling. When not needed, the Gear Wheel can be removed in one second, leaving a ratchet brace. This Brace is made of steel, and is heavily Nickel-plated, with Rosewood handle and Lignumvitæ head. The jaws are of forged steel, and will center and hold firmly round twist drills from $\frac{1}{8}$ to 7-16 of an inch in diameter. Also square shank Bits and Drills of all sizes. Also square and flat Screw-driver Bits. In fact, it will hold perfectly tool shanks of any size or shape. There is no other chuck in existence which will do this. It is our purpose to furnish everything in the line of Bit Braces and Breast and Hand Drills of a style and quality superior to anything else in the market.

With each Drill Brace we send an extra set of Forged Steel Jaws, mainly to be used for holding round twist drills less than one-eighth of an inch in diameter. As will be noticed in the cut, the large gear wheel has an extension handle to give it more power when needed.

Price of Drill Brace, complete...............per doz., $36 00
Same discount as Breast Drills.

BARBER'S IMPROVED PATENT BRACE.

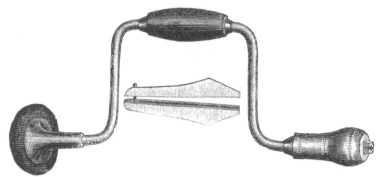

No. 21.—12-inch sweep...........................per doz., $11 00
" 22.—10 " " .... " 10 00
" 23.— 8 " " .... " 9 00
" 122.—10 " " Ratchet..................... " 18 00
" 123.— 8 " " " .... " 17 00

These Braces are intended to occupy a place midway between our highest and low-est price Grip Braces. They are made of Steel, polished, but not Nickel-plated. The heads and handles are Ebonized. The Jaws and Iron Quill are as seen in the Cut. They also have our new Anti-Friction Brass Collar. The Threads are all lathe cut and all parts of the brace are made for durability.

BARBER'S IMPROVED PATENT BRACE.

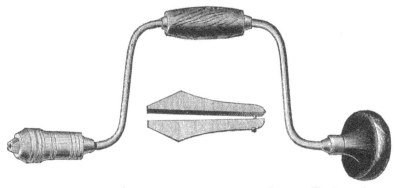

This Brace has the same style of Jaws as our best steel Braces. The Sweep is made of steel, tumbled and not polished; the head and revolving handle are of hard maple. It is a good, serviceable brace, and cheaper than any other brace of the same quality in market.

No. 71.—12-inch sweep...........................per doz., $8 00
" 72.—10 " " .... " 7 50
" 73.— 8 " " .... " 7 00
" 82.—10 " " Ratchet..................... " 14 00
" 83.— 8 " " " .... " 13 50

No. 92, 10 inch sweep.........$5 00 per doz. } Goodell's Self-cutting
" 93, 8 " " ...... 4 50 " } Thumb-screw.
" 103, 8 " " .... 2 75 " Plain Thumb-screw.

BARBER BRACE.

No. 50, 14 in. sweep, $20 00 doz.   No. 53, 8 in. sweep, $16 00 doz.
" 51, 12 " " 19 00 "   " 54, 6 " " 15 00 "
" 52, 10 " " 17 50 "

These Braces are well finished, and supplied with LIGNUMVITÆ HEADS and ROSE-WOOD HANDLES, and will hold all sizes of Square Shanks.

PATENT UNIVERSAL ANGULAR BIT STOCK.

The Universal Angular Bit Stock is presented to the public as a time economizer, to be used in connection with a brace and bit for boring holes in places where the brace and bit alone could not be used. As will be seen, it can be placed in the many angles or positions between the upper and lower braces (represented in the above figure). The ability to vary the angles, either at the commencement or during the operation of boring a hole, is an important feature.

Per dozen...$24 00

BARBER'S EXTENSION BIT HOLDER.

No. 1.

12, 15, 18, 21 and 24 inches.

Price, all lengths..............................per doz., $15 00

BARBER'S LATHE CHUCKS.

No. 2.

Price....................................... per doz., $18 00

DOUBLE EXTENSION WITH SCREW DRIVER.

Half length, 14 inches. Whole length, 21 inches.

Nickel-plated............................. per doz., $24 00

DRILL POINTS.

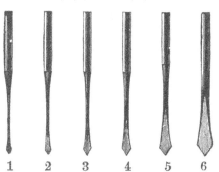

1 2 3 4 5 6

The cut represents the exact size of the drill points.

Price for set of six drills $0 25

HAND DRILL, No. 1.

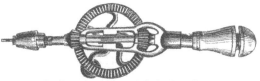

No. 1, Single Gear, hollow handle, nickel-plated.......per doz., $15 00
" 1 B, Double Gear, same chuck.................. " 18 00

Six drill points with each of the above drill stocks.

These drill stocks are made of malleable iron, with steel spindles and rose-wood head and handle. The jaws are of forged steel, and will hold perfectly any sized drills named below.

They are the only drill chucks in use which will hold Morse Twist Drills from 1-32 to ⅛ inch.

HAND DRILL, No. 2.

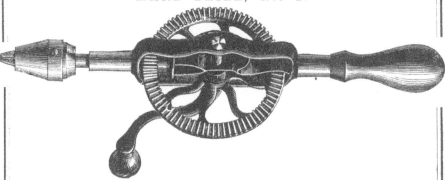

The Chuck of this Drill Stock is the same style as the No. 1, but it will hold ¼ inch drills and all smaller sizes. It has cut gears, is heavily nickel-plated, with rose-wood head and handle. The head is hollow and contains six drill points. It is a tool much in demand.

Price..per dozen, $30 00

HAND DRILL, No. 3.

The No. 3 is same size, style and finish as the No. 1, but with a longer crank and a hollow handle long enough to hold Twist Drills.

Price, No. 3 ....per dozen, $18 00
" " 3 B, Double Gear..................... " 21 00

HAND DRILL, No. 4.

This Drill Stock is eight inches in length and weighs eight ounces. It is made of iron with rose-wood handle, and brass chuck for holding the drill points. This chuck is made on a new plan, and it centers and holds the drill perfectly. With each drill stock we send a box containing six superior drill points, of various sizes.

Price of Stock and Drills..............................per dozen, $6 00

JOHNSON'S AUTOMATIC BORING TOOLS.

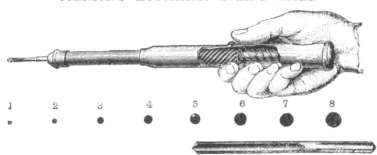

This instrument is designed for boring wood for various purposes, such as for setting brads, finishing nails, screws, etc., eight bits (or drills, size indicated by dots and numerals above) accompanying each tool. It can be used in many places where the bit-brace, gimlet, or brad-awl cannot, and is superior to either for the purposes mentioned. Piano-forte and organ builders, as well as cabinet-makers and house carpenters, who have used them, all attest their great value as a tool for doing their work rapidly and perfectly.

Price, with Drills..................................per dozen, $24 00

Points......................................   "   72

   "   ....per set,   48

GRAVES' AUTOMATIC DRILL STOCK AND DRILL POINTS.

N. 1886

This is a most complete tool for bracket saw work. The stock is made of rosewood, with steel spindle.

The drill points are easily adjusted by means of a brass thimble, which screws on to the end of the spindle. It is easily worked with one hand, and will drill the most delicate wood without danger of splitting.

Price, with one dozen Drill Points assorted, in wood box... $ 1 00

Per dozen.. 12 00

BREAST DRILL, No. 10.

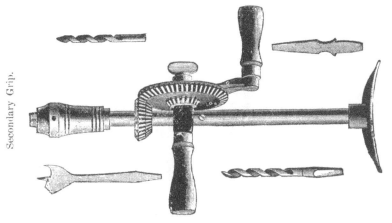

Secondary Grip.

This Drill is made of round wrought iron, $\frac{5}{8}$ of an inch in diameter.   The handles are rose-wood, the head malleable iron, and the chuck jaws of steel.   It has a changeable gear, one even, and the other speeded three to one.   The change from one to the other can be made in one second.   The chuck will hold any shape shank, round, square, or flat, as seen in the cut.   These tools are not sold with the drill stock, but are only put in to show the shape of shank which the chuck will hold.

This Drill Stock is heavily Nickel-plated, and presents a very beautiful appearance. It would be very attractive in a show window.   This drill stock has cut gears.   An extra set of steel jaws goes with each drill stock, for holding small round drills.

Price...per doz., $36 00

BREAST DRILL, No. 11.

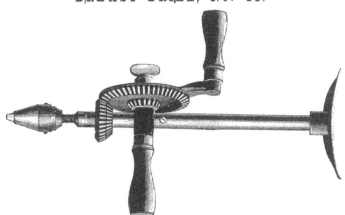

This Drill Stock is the same as No. 10, but the chuck will only hold round shank drills.   Being limited in its range, it holds drills from 1-32 to $\frac{1}{4}$ inch very firmly and true, and is a most perfect tool, heavily Nickel-plated, with rose-wood handles.   The gears are cut.

Price.................................   ....per doz., $36 00

We are now putting a Secondary Grip on our No. 10, 12 and 13 Breast Drills which works automatically and will hold round twist drills perfectly.   It adds much to the cost of goods, but nothing to the selling price.

BREAST DRILL, No. 12.

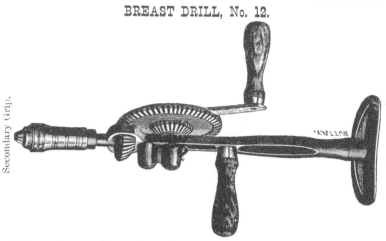

This is the Drill which we have made for many years and sold at $30 list. It has a Malleable Iron Stock, Japanned, Rosewood Handles, Polished and Plated Chuck, changeable Gears, one even and the other 3 to 1. It has a Barber Improved Chuck with our recent improvement, which makes it hold perfectly tools of all shapes and sizes.

Price . . . . per doz., $30 00

BREAST DRILL, No. 13.

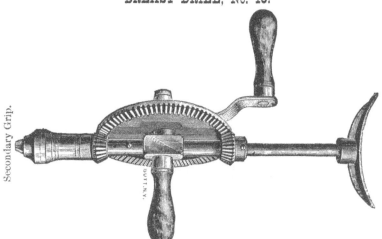

This is the largest sized Drill in market, the drive-wheel being 6 inches in diameter, giving a speed of 4½ to 1. It is double-geared and most perfect in every part. It has Cut Gears, Steel Stock, Rosewood Handles, Steel Jawed Chuck, which will hold any size square and round tool shanks. It is heavily Nickel-plated, and really the most expensive Breast Drill in the market.

Price . . . . per doz., $48 00

Both Drill Stocks on this page have extra steel jaws for holding small round drills.

PRICE OF UNIFORM SHANK DRILLS.

Diameter.	Length.	Price each.	Diameter.	Length.	Price each.
1-8	5 1-8 in.	$0 45	7-16	7 1-4 in.	$0 90
5-32	5 3-8 "	45	15-32	7 1-2 "	95
3-16	5 5-8 "	50	1-2	7 3-4 "	1 00
7-32	5 7-8 "	55	17-32	8 "	1 10
1-4	6 1-8 "	60	9-16	8 1-4 "	1 20
9-32	6 1-4 "	65	19-32	8 1-2 "	1 30
5-16	6 3-8 "	70	5-8	8 3-4 "	1 40
11-32	6 1-2 "	75	21-32	9 "	1 50
3-8	6 3-4 "	80	11-16	9 1-4 "	1 60
13-32	7 "	85	3-4	9 3-4 "	1 85

The Shanks to the above Drills are 2¼ inches long and ½ inch in diameter, to fit hole in spindle to *Anvil Drill*, and Drilling machines.

We also have in stock STRAIGHT SHANK DRILLS, from 1-16 to 1-4 inch, which can be used in the Chuck accompanying the Anvil Drill, and Drilling machines.

STRAIGHT SHANK DRILLS.

Diameter.	Length.	Price per dozen.	Price Each.	Diameter.	Length.	Price per Dozen.	Price Each.
1-16	2½	$1 00	$0 9	19-64	4⅜	$3 90	$0 35
5-64	2⅝	1 10	10	5-16	4½	4 20	37
3-32	2¾	1 20	11	21-64	4⅝	4, 50	40
7-64	2⅞	1 30	12	11-32	4¾	4 80	42
1-8	3	1 45	13	23-64	4⅞	5 10	45
9-64	3⅛	1 60	15	3-8	5	5 40	48
5-32	3¼	1 80	16	25-64	5⅛	5 70	50
11-64	3⅜	2 00	18	13-32	5¼	6 00	53
3-16	3½	2 20	20	27-64	5⅜	6 40	55
13-64	3⅝	2 40	21	7-16	5½	6 80	59
7-32	3¾	2 65	23	29-64	5⅝	7 20	63
15-64	3⅞	2 90	26	15-32	5¾	7 50	65
1-4	4	3 15	28	31-64	5⅞	7 75	67
17-64	4⅛	3 40	30	1-2	6	8 00	70
9-32	4¼	3 65	32				

BIT STOCK DRILLS.

FOR METAL OR WOOD.

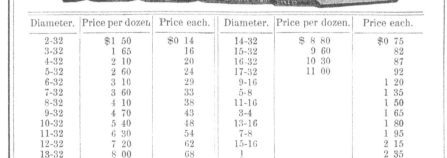

Diameter.	Price per dozen	Price each.	Diameter.	Price per dozen.	Price each.
2-32	$1 50	$0 14	14-32	$ 8 80	$0 75
3-32	1 65	16	15-32	9 60	82
4-32	2 10	20	16-32	10 30	87
5-32	2 60	24	17-32	11 00	92
6-32	3 10	29	9-16		1 20
7-32	3 60	33	5-8		1 35
8-32	4 10	38	11-16		1 50
9-32	4 70	43	3-4		1 65
10-32	5 40	48	13-16		1 80
11-32	6 30	54	7-8		1 95
12-32	7 20	62	15-16		2 15
13-32	8 00	68	1		2 35

Our Bit Stock Drills will fit any brace in the market, and will drill Steel, Iron, or other Metals, as well as Wood. They are not injured by contact with screws or nails, and will bore straight any kind of Wood without splitting it.

ANVIL, VISE AND DRILL.

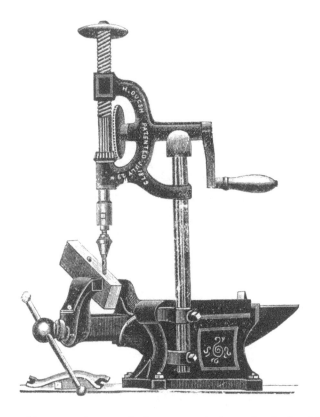

This machine was first made by a practical mechanic for his own use, as he could find nothing in the market that would answer his purpose. When finished, so many other mechanics wanted it that he was minded to get it patented and put it on sale. When brought to our notice, we saw at once its great utility, and bought the patent for the whole United States.. The anvil has a steel face, 4x8 inches. Vise Jaws 3½ inches wide, and steel faced. Steel Drill Press with adjustable chuck to hold ¼ inch drills and all smaller sizes. We have for sale Morse Twist Drills from ¼ to ¾ inches, with uniform ½ inch shanks, as seen in the cut opposite, which just fits into the spindle when the chuck is removed; one of these drills, for a sample, goes with each machine. See list of prices of Drills, page 11.

VISE AND DRILL.

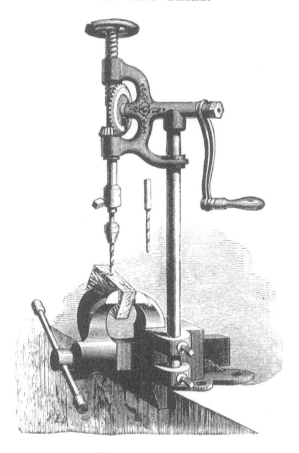

The article to be drilled can be held firmly in the vise, so as to be drilled at an angle. A bolt of iron or soft steel can be drilled accurately down through the centre, which cannot be done by hand with any other tool. When the Drill is sold without the Anvil, it is attached to an Offset Vise, as seen in the right hand cut. This Vise has steel-faced jaws 3½ inches wide, which entirely overhang the screw and the box which covers it. For many purposes this is the most convenient Vise in use.

Weight of Anvil, Vise and Drill			80 pounds.	Price,	$18	00
" " "	and Vise		60 "	"	10	00
"	" Drill Press		20 "	"	8	00
"	" Offset Vise and Drill		61 "	"	15	00
" " "	" alone		41 "	"	7	00

MILLERS FALLS VISE.   PIPE VISE.

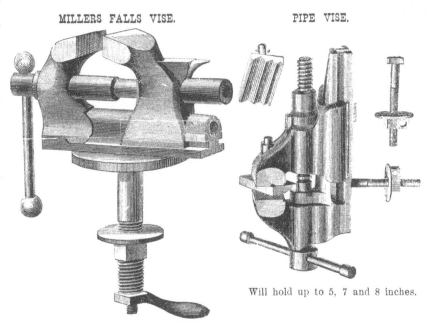

Will hold up to 5, 7 and 8 inches.

MILLERS FALLS VISE.

Union and Backus Vise Combined.--Heavy or Chipping Vise, Covered Screw.

If Seats are wanted, please so mention in your orders. Prices at Millers Falls, Mass., or 74 Chambers St., New York.

Width of Jaw.	Without Iron Seats.	With Iron Seats. EXTRA.	Weight.	Width of Jaw.	Without Iron Seats.	With Iron Seats EXTRA.	Weight.
1½ in.	$3 00	$0 50	3½ lbs.	4 in.	$10 00	$1 25	56 lbs.
1¾ "	3 50	0 50	5 "	4½ "	12 50	1 50	67 "
2 "	4 00	0 50	9 "	5 "	17 00	1 75	75 "
2½ "	5 50	0 75	22 "	6 "	25 00	2 00	116 "
3 "	7 00	1 00	39 "	7 "	30 00	No Seat	160 "
3½ "	9 00	1 00	47 "				

PIPE VISE.

Width of Jaw, 4 inches	Weight, 57 lbs	Price, $17 50
" " 5 "	" 79 "	" 23 00
" " 8 "	" 153 "	" 33 00

" Pipe " Vise, showing extra jaw in place. The extra jaw has corrugated grooves of different sizes, by means of which pipe of any dimensions may be firmly held without injury, and without the necessity of screwing up the Vise excessively tight. The extra jaw may be removed in a moment when the Vise is wanted for other work (see cut). This Vise is made extra heavy and strong with very heavy jaws. (see cut).

The 8 inch allows the extra jaw to be used on either side, so as to hold large pipe perpendicularly. The 4 and 8 inch have covered screws.

A bolt for securing the Vise to the bench, with patent nut, is included in all the sizes without extra charge.

HEAVY CHIPPING VISE.

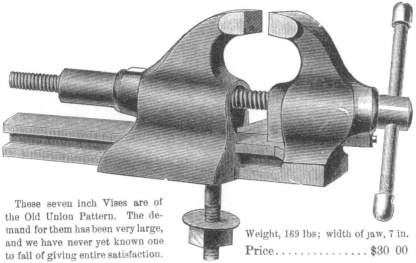

These seven inch Vises are of the Old Union Pattern. The demand for them has been very large, and we have never yet known one to fail of giving entire satisfaction.

Weight, 169 lbs; width of jaw, 7 in.
Price............... $30 00

FINISHING VISE.

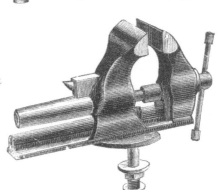

Width of Jaw.	Weight,	Price.	With Seat Extra.
3 inch	24 lbs	$6 50	$1 00
4 "	42 lbs	9 00	1 25
5 "	60 lbs	14 50	1 75

OVAL SLIDE COACH VISE.

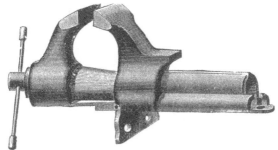

Width of jaw, 4 inches; Depth of Jaw, 2 inches; Opens 9 inches; Depth from screw cover to top of jaw, 5½ inches; weighs 43 lbs.

Price................. .. .. each, $8 00

MECHANICS' VISE.

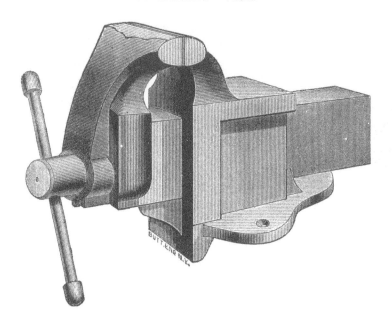

Width of Jaw.	Opens.	Weight.	Price, each.
2 inch............3½ inches.........		8 pounds............	$4 00
2½ " ....3½ "		18 "	5 50
3 " ....4½ "		32 "	7 00
3½ " ....4½ "		42 "	9 00
4 " ....5 "		50 "	10 00
4½ " ....5 "		60 "	12 50
5 " ....6 "		76 "	17 00

This is probably the best square Box Vise in market, being first-class in every particular. The Jaws are steel-faced, and the Anvil is also faced with steel, which we believe is not the case with any other Vise in use.

OVAL SLIDE PARALLEL VISE.

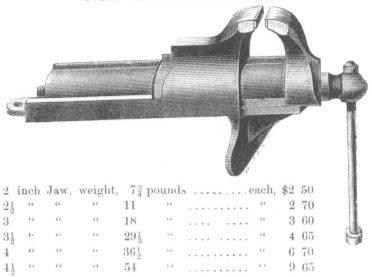

2 inch Jaw,	weight,	$7\frac{3}{4}$ pounds	.... each,	$2 50
$2\frac{1}{2}$ "	"	"	11	"	.... "	2 70
3 "	"	"	18	"	.... "	3 60
$3\frac{1}{2}$ "	"	"	$29\frac{1}{3}$	"	.... "	4 65
4 "	"	"	$36\frac{1}{2}$	"	.... "	6 70
$4\frac{1}{2}$ "	"	"	54	"	.... "	9 65

OVAL SLIDE VISE.

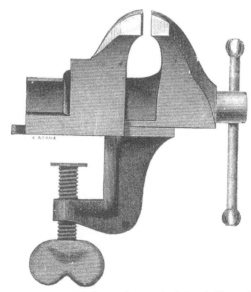

This Vise has a clamp by which it can be attached to a table, and can be removed at pleasure by turning a Thumb Screw.

$1\frac{1}{4}$ width of Jaw	$1 25
$1\frac{3}{4}$ " "	1 50

JACK-SCREWS. BELL BOTTOM.

	Diam. of Screw.	Height of Barrel.	Height of Jack when turned down to the lowest point.	Net rise.	Whole Height.	Weight.	Price.
No. 1.	1¼ in..	6 in....	8 in....	4 in....	12 in....	10 lbs..	$ 2 50
" 2.	1¼ in ..	7 in ...	10 in....	6 in ...	16 in....	11¾ lbs..	3 00
" 3.	1½ in...	7½ in ...	10 in ...	5 in ...	15 in ..	18½ lbs..	3 25
" 4.	1½ in...	9 in ...	12 in...	7 in ...	19 in ...	18 lbs .	3 75
" 5.	1½ in...	10½ in	14 in....	9 in ...	23 in....	25 lbs....	4 00
" 6.	1¾ in...	9 in....	12 in	6 in....	18 in....	24 lbs....	4 25
" 7.	1¾ in...	10½ in ...	14 in....	8 in....	22 in ...	28¾ lbs..	4 50
" 8.	1¾ in...	12½ in ...	16 in....	10 in....	26 in ...	33½ lbs..	5 00
" 9.	1¾ in...	15 in....	18 in ...	12 in....	30 in....	37½ lbs..	5 50
" 10.	2 in...	8½ in	12 in...	5 in....	17 in....	31 lbs..	5 50
" 11.	2 in...	10½ in....	14 in....	7 in....	21 in....	36 lbs..	6 00
" 12.	2 in..	12 in ...	16 in....	9 in....	25 in....	41½ lbs..	7 00
" 13.	2 in...	16 in ...	20 in....	13 in ...	33 in....	50 lbs..	8 00
" 14.	2½ in...	10 in ...	14 in....	8 in ...	22 in....	48 lbs..	8 50
" 15.	2½ in...	12 in ...	16 in....	10 in ..	26 in....	53¼ lbs..	9 50
" 16.	2½ in...	15½ in ...	20 in....	14 in....	34 in....	69 lbs..	11 00
" 17.	2½ in...	20 in ...	24 in....	18 in....	42 in....	85 lbs..	13 00

CAR JACKS.   (See Cut on next page.)

	Diam. of Screw.	Height of Barrel	Height of Jack when turned down to the lowest point.	Net rise.	Whole Height.	Weight.	Price.
No. 18.	2 in...	8½ in ...	12 in	5 in....	17 in....	31 lbs....	$5 50
" 19.	2 in...	10½ in....	14 in.....	7 in....	21 in ...	36 lbs....	6 00
" 20.	2 in...	12 in	16 in.....	9 in ...	25 in....	41 lbs....	7 00

These Jacks have cast iron barrels, with wrought iron screws cut in a lathe. There is no better Jack made.

Order by number and there will be no mistake.

CAR JACK.

HERCULES IRON CUTTER.

During the past ten years many attempts have been made to produce a good iron cutter at a small cost. Prices went so low that we could not meet them, and so withdrew our Hercules Cutter from the market. Time has proved that a cheap and light machine will not do the work for any great length of time, and so all that class of cutters have disappeared. We have added still more to the weight of the Hercules Cutters, and now offer them in the full confidence that they will not break and will do the work as underneath stated. Every iron merchant should have one for his own use and dispense with the slow and expensive use of cold chisels. It is also very useful in all ship yards, factories, machine shops and large blacksmith shops.

No.	Weight.	Cuts.	Price.
1	62 lbs.	$\frac{3}{8}$x2 in. or $\frac{1}{2}$ in. round or square	$30 00
2	165 "	$\frac{5}{8}$x2 " $\frac{3}{4}$ " "	45 00
3	400 "	$\frac{7}{8}$x4 " 1 " "	60 00

Every machine cuts larger iron than here stated before it leaves our factory.

LESTER IMPROVED SAW.

This machine is adapted alike to the use of either *amateurs* or *mechanics*, and will be found very convenient in *pattern making*.

The Lester Saw is made of iron and steel, except the Pitman and Arms, which are ash. The *Iron work* is nicely japanned black, with stripes of red and gilt. The *Wood work* is varnished, bringing out the beautiful grain of the ash.

The *Driving-wheel* is 15 inches in diameter, and weighs 13 pounds. The *Treadle* is made of *Iron*, from a new and handsome pattern.

The *Scroll Sawing Attachment* is provided with a *Tilting Table* for inlaid work. The arms swinging 18 inches in the clear. A Dust Blower which works automatically. A Roller inserted in the Table back of the saw, which makes it run perfectly true. The clamps will hold the coarsest or finest saw perfectly, while they are made adjustable to the right or left, or backwards and forwards, so that the blade may be kept in line. This saw will cut lumber from one-sixteenth inch to two inches in thickness.

The *Circular Saw Attachment* consists of a saw three inches in diameter, and an iron table, three by four and one-half inches. This saw will cut lumber up to one-half inch.

The machine is also furnished with a *Solid Emery Wheel* and *Drilling Attachment*, accompanied by six best Stubbs Steel Drills, of assorted sizes.

The *Lathe Attachment* is provided with iron ways and rests, steel centers, and three fine steel turning tools. Length of bed, 15 inches. Distance between centres, 9 inches. Swing, 3 inches. Length of slide rest, 4 inches.

The following parts are supplied with each machine, *viz.:* Six Saw Blades, One Wrench, Screw Driver, Two Sheets of Designs, and a nice box for the small tools.

When shipped it is taken apart and packed in a box, but can easily be set up by following the directions accompanying each machine.

Price for the above, with Nickel-plated Table and Saw Guide..............$10 00
Price for the Saw only, without Lathe and Circular Saw, Nickel-plated Table.   8 00
When desired we furnish with the Lathe a very nice Drill Chuck for working
    metals and a Tail Stock with Screw Center, for....................Extra   2 00
We furnish the Scroll Saw Guide alone for...............................   25
We also furnish a larger *Table for Circular Saw* 6 by 7 inches, provided with
    a guide for sawing straight edges, for............................Extra   50

We have recently invented one of the greatest improvements which has been applied to saws for many years. It is a clamp and strainer all in one. By simply moving a lever at the front end of the upper arm the saw blade is clamped and strained instantly.

The roller which we now insert in the Table at the back of the blade makes our $10.00 Saw do just as good work as a Saw which runs in guides and costs $20.00. It not only does as good work, but runs much easier. Our new Rubber Blower is by far the best one now in use.

We have made many other smaller improvements which will be highly appreciated.

THE NEW ROGERS SAW.

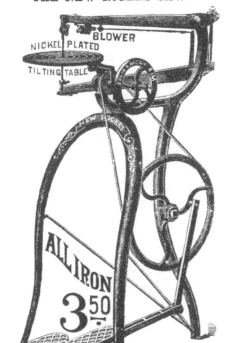

BLOWER
NICKEL PLATED
TILTING TABLE
NEW ROGERS
ALL IRON
3.50
J. S. CONANT SC.

The latest improvements on this Saw are Pratt's Rubber Positive Blower, a new Clamp for holding the Saw Blades, and a Roller inserted in the Table back of the Saw, which makes it run perfectly true.

This season we have added all the improvements that our ingenuity can suggest, and have no hesitancy in declaring our New Rogers to be the BEST CHEAP SAW offered to the public.

The entire frame-work is made from Iron, painted and japanned black, and ornamented with red and gilt stripes.

The arbors, etc., are made of steel, and are carefully gauged and fitted to their bearings.

The Arms and Pitman are of the best *Selected Ash*.

ALL PARTS MADE TO INTERCHANGE.

While our Jig Saw possesses *every improvement* and *advantage claimed* for any other in the market, we further claim the following points of superiority, *viz.:*

1. The *Bearings to the Arms* are carefully sized to bring them in perfect line. (This is a vital point in the construction of any Jig Saw.)

2. We provide each machine with a *Dust Blower*, which is a very *great advantage.*

3. Our machine has a *jointed* Stretcher Rod, which allows the operator to throw the upper arm out of the way when adjusting his work or saw. This joint also permits the machine to work much more freely than with a straight iron rod.

Price of No. 2 Rogers Saw, - - - $4.00

4. Our clamps have a *hinged jaw,* which overcomes the disagreeable raking overthrow of the blade, which is unavoidable when the saws are secured rigidly to the arms. Saw blades are not nearly so liable to

Price of No. 1 Rogers Saw, with Balance of *Iron* instead of *Emery*, and with *Tilting Table Japanned* instead of Nickel-plated, - - 3.50

break when clamps have this joint. Thus a *large percentage* of the *expense* of running the saw is saved. Besides this the saw runs *much easier*, the swing coming at the *hinge* instead of bending the blade at each stroke of the saw.

5. The Balance Wheel is 4¼ inches in diameter, with a handsome spoke center and *Rim* of *Solid Emery* (No. 70) 5-8 x 3-4.

6. Our attachment for *Drilling* is on the *Right Hand* side of the machine, which, for convenience, is an obvious advantage.

7. We use *no Pins* in the construction of our machine, but prefer the durability of nicely fitted *screws* and *bolts* in securing each part.

8. Each machine is *set up* and *run* and *carefully inspected* before leaving our works. It is then taken down and shipped in boxes.

These points of superiority are neither possessed by nor claimed for any other Cheap Saw Our Saw is provided with a *polished Tilting Table*, heavily Nickel-plated.

While the NEW ROGERS SAW is very rich, though not gaudy in appearance, it has been more especially our object to make, for the least possible money, a saw characterized for its Compactness, Strength and Durability, ease of action, and firmness when in operation.

With each machine we give Six Saw Blades, Wrench, Sheet of Designs and Three Drill Points. The Saw alone weighs 25 lbs.; Saw and Box together, 36 lbs.

GOODELL LATHE.

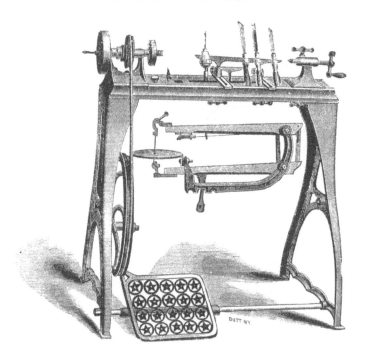

This lathe was designed by Mr. A. D. Goodell, for many years the master mechanic at our factory in Millers Falls, and the inventor of many of our most valuable machines. It is patterned almost exactly after the latest improved lathes now used in the best machine-shops and pattern-makers' rooms. The large driving-wheel has two grooves, of varying depths, on its face, to give a change of speed, as the belt runs from it to the cone pulley on the lathe head. The lathe head is provided with a two-inch face-plate, a spur center, a screw center for turning cups, and also with a very nice drill chuck to hold from 1–32 to 1–4 inch round twist drills for drilling wood or iron. It also has on the outer side of the pulley a 4½ x ⅞ inch solid emery wheel, and a drill spindle, with set screw, to hold drill points for wood drilling. The tail stock has a screw feed center, which is secured at any point desired, by a clamping arrangement like that on the modern engine lathes. It is also provided with a long and short tool rest, five turning tools, wrench, drill points, etc., etc. Swing of lathe, five inches. Length of bed, 24 inches. Distance between centers, 15½ inches.

The Scroll Sawing attachment, as seen in the cut, is secured to the lathe bed by one bolt, and can be put on or off at pleasure. This attachment has all of the improvements found in any of our scroll saws.

This lathe is thoroughly built and highly finished, the planed and polished parts being Nickel-plated. Weight 56 lbs.

Price of Lathe and Lathe Tools..... ....$10 00

    "   Scroll Saw Attachment...................... .... 2 00

In ordering please state whether the Saw attachment is wanted.

COMPANION SAW.

We are making this machine especially for the *Youth's Companion.*

It is fully illustrated in their Premium List. The publishers of this paper were the first to inaugurate scroll sawing in this country, and it has been through their efforts mainly that the business has had such great popularity.

They have learned exactly what kind of a machine is most wanted, and recently planned this one to meet the demand. The lathe is large enough to do good work, and is supplied with all the tools and instructions needed for putting it into successful use.

The Lathe Bed is 24 inches long, 15 inches between centers, and 5 inches swing. The Scroll Saw can be taken off by loosening one bolt. The quality and price of this Companion Saw will ensure it a very large sale.

Price, Complete............................. ....$8 50
   "   of Lathe and Tools 7 00

FAMILY TOOL CHEST.

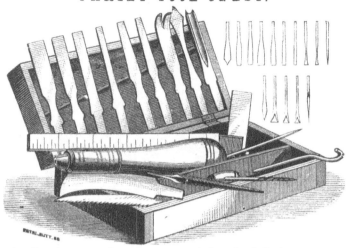

No. 1.—Contains a complete assortment of Cast Steel Tools (20 in number); also a patent Adjustable Holder, with an additional number of 12 Brad-awls, Chisels, etc., and all enclosed in a nice Walnut Case.

Per dozen... ....$42 00

No. 2.—Contains an assortment of Cast Steel Tools (9 in number); also a Patent Adjustable Holder, with an additional number of Brad-awls, Chisels, etc., and all enclosed in a neat Walnut Case. Especially adapted to the wants of every family.

Per dozen..$20 00

CRICKET SAW.

TILTING TABLE.

ALL IRON

To meet the demand for a cheap and good Foot Power Saw, we have made an all Iron Machine which we offer for $2.50. For Scroll Sawing alone it is about as good as the higher price Saws and the finish is the same as our Lester and Rogers Saws. It has an Iron Tilting Table, Second growth Ash Arms, eighteen inches long, improved Clamps, &c. Weight of Saw, seventeen pounds; weight when boxed, thirty-five pounds. Boxed for shipping without extra charge.

Price...................... .... .$2 50

LESTER SAW REPAIRS.

Arms, per pair..............	$0 30
Arm Bolts and Nuts........	6
Balance Wheel............	50
Belt	15
Brace...................	10
Circular Saw............	25
" " Arbor........	15
" " Table.........	12
Clamps, per pair........	35
Drills, per doz............	50
Drill Screws, each..........	2
Drive Wheel Arbor........	15
Drive Wheel..............	1 25
Emery "	40
Frame	1 50
Front Leg...............	85
Lathe Bed...............	35
" Head Stock.........	25
" Tail "	16
" Rest...............	5
" " Socket.........	10
" Pulley............	10
" Tail Centre.........	10
" Spur..............	20
Lower Arm Guide.........	5
Locking Bolts.............	3
Pitman, Long	10
" Short.............	5
Pulley...................	10
Pitman Screws............	5
Rest Socket Bolt..........	5
Screw Driver.............	25
Scroll Saw Table, Nickel....	65
" " " Japanned..	30
" " " Guide....	5
Spindle and Crank.........	20
Stretcher..	15
Stove Bolts, each..........	2
Treadle Caps, per pair......	10

Treadle..................	$0 50
Tool Box.................	12
Blower...................	25
Table Bolts..............	3
Thumb Nuts.............	5
Thumb Screws...........	5
Turning tools.............	1 00
Wrench....	6

BOGERS SAW REPAIRS.

Arm Bolt....	$0 2
Arm Castings.............	40
Arms, per pair...........	30
Back Leg.................	30
Balance Wheel, Iron........	25
" " Emery......	40
Belt.....................	15
Brace.......	15
Blower	20
Crank and Spindle.........	20
Crank Screw..............	5
Clamps, per pair..........	25
Drill Screw..............	2
Drive Wheel Bolt and Nut..	10
Drive Wheel..............	50
Front Leg.................	60
Headless Screws..	2
Locking Bolt and Nut......	5
Pitman...................	5
" Screw.............	5
Pulley...................	10
Stretcher................	15
Stove Bolt...	2
Table, Nickel Plate........	60
" Japanned	25
Table Bolt........	3
Treadle.................	25
Wrench..................	6

No. 2. BRACKET SETS.

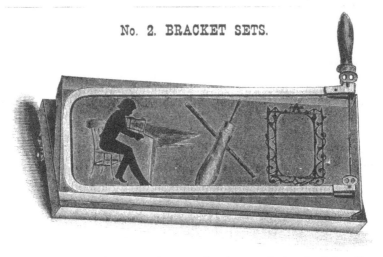

These sets are packed in paper-boxes, and embrace a Spring Steel Saw Frame, polished and nickel-plated, 5 x 12 inches, having patent Clamps for holding the Saw blades, rose-wood handle. Fifty full sized designs, embracing a great variety of fancy and useful articles; six Saw Blades, one awl, one Sheet of Impression Paper, and full directions for using the Saw.

Price.................................per dozen, $15 00

No. 1. Pleasure and Profit Set, frame rough nickeled, with Japanned handle, box same size and same contents as No. 2.
Price..................................per dozen, 10 00

PRICE OF SAW PARTS.

	Companion.	Goodell.
Lathe Bed...	$ 75	75
Legs. Each...(2)	75	75
Treadles...	50	50
" Rod...	15	20
" Caps. Each.............................(2)	5	5
Driving Wheel	1 25	1 25
" " Bolt.......................................	15	15
Emery "...	40	40
Head Block	20	25
" " Spindle and Face Plate...................	25	25
" " " Cone Pulley................	15	20
Tail "...	20	25
Tail Block Spindle and Crank................................	20	30
" " " Binder..............................	10	15
Long Rest...	20	30
Short "...	10	15
" Sockets. Each............................(2)	15	15
Spur Centre..	20	20
Screw " for Cups.......................................	15	15
Star Chucks..	....	2 00
Saw Attachment Arm Casting.	50	50
Arms..per pair.	30	30
Table..	60	25
Stretcher..	15	15
Clamps ....per pair.	35	35
Bolts...	3	5
Turning Tools...(5)	....	1 25
" "...(3)	75	....
Arm Bolts and Nuts..	6	6
Belt..	15	15
Drill..per doz.	50	50
Drill Screws.......................................each.	2	2
Pitman, Long..	10	10
" Short..	5	5
Pitman Screws...	5	5
Screw Driver..	25	25
Blower	25	25
Table Bolt..	3	3
Wrench...	6	6

PRICE LIST OF FANCY WOODS.

PLANED TO FOLLOWING THICKNESS.	⅛ inch.	3-16 inch.	¼ inch.
Black Walnut.................per foot.	8c.	9c.	10c.
White Holly.................... "	9	10	11
Poplar........................ "	4	5	6
Plain White Maple .... "	5	6	8
Oak.......................... "	7	8	10
Cherry....................... "	7	8	10
Red Cedar .... "	8	10	12
Spanish Cedar..... .... "	8	10	12
Mahogany.................... "	10	12	14
Bird's-eye Maple .... "	12	15	16
Cocobola.................... "	16	18	20
Rose-wood... .. .. "	16	20	25
Amaranth. .... "	16	20	25
Satin Wood.................. "	20	25	30
Hungarian Ash............... "	30	40	50
Tulip........................ "	30	40	50
Ebony....................... "	40	50	60

PATENT BRACKET SAW FRAMES.

WOOD HIGHLY POLISHED.

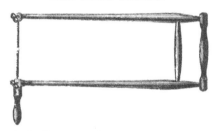

Prices 12 and 14 inch Rose-wood...................per doz., $9 00
" 12 " 15 " Birch, ... " 6 00

SPRING STEEL.

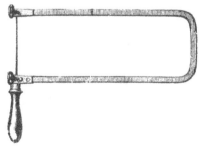

8 inch.................................per dozen, $13 20
10 " .... " 14 40
12 " .... " 15 60
14 " .... " 16 80
16 " .... " 18 00

BRACKET SAW BLADES.

We have dealt in all kinds of imported and American Blades, but our trade has gradually settled down to the Star Bracket Saw Blade until at present hardly any other kinds are wanted. We not only supply this country but export large quantities to England, France and Germany. We are free to say that these goods are much the best in use.

THE STAR SAW BLADES.

Per Gross, Nos. 000 to 6					$1 00
"	"	"	7		1 10
"	"	"	8		1 20
"	"	"	9		1 30
"	"	"	10		1 40
"	"	"	11		1 50
"	"	"	12		1 60

CARVING TOOLS.

Six Tools in a Wooden Box.

Price, per dozen sets $12 00

These Tools are forged from the best quality of steel, and are fully warranted. They have Rose-wood Handles and are sharpened ready for use.

CLOCK MOVEMENTS.

WITHOUT CASES.

This Cut Represents Russell's Design, No. 217. Movement No. 4 fits this design.

No. 1. Marine (Spring) movement, with 2½ inch (Nickel) Sash and Dial, complete, runs thirty hours................each, $1 50

No. 2. Pendulum movement, with 3 inch Nickel-plated Sash and Dial, runs eight days.............................each, 2 00

No. 3. Marine movement, with 3 inch Nickel-plated Sash and Dial, runs eight days.............................each, 2 00

No. 4. Pendulum movement, with 5 inch Nickel-plated Sash and Dial, strikes hours and half hours, runs eight days, each.. 3 00

No. 5.

Something New and Novel,

Demanding the attention of every scroll sawyer, and which is now presented to the public for the first time in this country, having hitherto been made principally by the Swiss and Germans. **A CUCKOO CLOCK MOVEMENT**, made from Russell's new design, No. 218, as shown in the annexed cut, in which the Cuckoo appears at the little door over the dial and announces the hour in a flute-like voice, similar to the cuckoo's song. If the hour hand indicates *one* the door opens, the bird appears, bows and sings "cuckoo" *once* and disappears; if the hand is at *two*, he sings "cuckoo" *twice* and so on up to *twelve* when he sings *twelve* times. This movement is not so complicated as it is supposed to be, any one who can put in an ordinary movement can case this one, the extra attachments being all in place when the movements are sent. This clock is an immense source of amusement both to children and older people, and will amply repay the sawyer for his outlay, aside from the fact that the same movement in the same case would bring $40 or $50 in New York. This case may be made as an Imitation Cuckoo Clock, and other movements put in.

Price of movement (which is an 8-day pendulum), with 5-in. Nickel Dial, complete, $8 15
   "   "   " "   "   "   Brass   "   "   8 00
   "   "   " "   "   without Dial, but with Ivory

Hands and Figures.. 8 25

Moulding sufficient to trim case... 25

ALFORD HAND VISE.

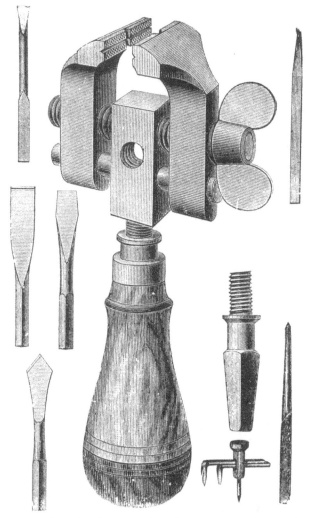

The jaws are of forged and tempered steel, the screw and cross bar are also made of steel, the handle is made of rose-wood with lignumvitæ cap. It is hollow, and the bit shank and tools seen in the cut are placed inside. The blades bent at right angle, are used for cutting washers.

The Vice Jaws are 1¼ inches wide and open 1⅛ inches. They will centre and hold tools firmly of any shape. The vise is one-third larger than the cut, while the tools are full size. The handle can be unscrewed from the vise and the bit shank put in its place, to be used with a bit brace for any kind of boring. drilling or cutting washers. The handle can also be screwed into the vise at right angles with its usual position, which is desirable for many kinds of work.

Price, with all the tools................. ....each, $1 75

STAR CHUCK.

No. 1, with Shank $\frac{1}{2}$ inch in diameter, and two inches long, holds Drills from 1-64 to $\frac{1}{8}$ inch. Price, $1 25 each.

No. 2, with shank $\frac{5}{8}$ inch in diameter and $3\frac{1}{2}$ inches long, holds Drills from 1-32 to $\frac{1}{4}$ inch. Price, $2 00 each.

These Chucks are simple in construction, strong and durable, and are warranted to hold the Morse Twist Drills firmly, without danger of breaking. The Shanks (or Mandrels) are centered so they can easily be fitted to any lathe desired. The above cut represents No. 1, full size (except in length of shank).

LEWIS' PATENT SINGLE TWIST SPUR BITS.

Size. Sixteenths.	Auger.	Car.	Size. Sixteenths.	Auger.	Car.
4	$3 50	$7 00	13	$7 50	$15 00
5	4 00	8 00	14	8 00	16 00
6	4 00	8 00	15	9 00	18 00
7	4 50	9 00	16	10 00	20 00
8	5 00	10 00	17		23 00
9	5 50	11 00			
10	6 00	12 00	Size.		
11	6 50	13 00	32½ qrs.,	$6 60	$13 20
12	7 00	14 00	24 "	5 50	11 00

GERMAN BITS.

4 to 12-32 . . . . $1 00 per dozen.

PRATT'S PATENT ELASTIC BOILER TUBE SCRAPER.

We say this is the leading Tube Scraper in the market, and the best one. If you order them, and find out that this is not true, return with charges for cost and trouble.

The price is strictly **ONE DOLLLAR PER INCH.** We will allow the Hardware Trade a large discount, so as to make it an object for them to introduce it.

PORTABLE CAMP STOOL.

Made of second growth ash.   Weight, one pound.

Price, $6 00 per dozen.

PRATT'S PATENT AUGER HANDLE.

So far as we are aware this is the only Auger Handle in market which will centre perfectly and hold Auger Shanks of all shapes and sizes.   Only one handle is needed as its range is from the largest to the smallest size Auger in use.   A steel band, 3 inches long and 3-16 thick, is made to fit the handle.   This band is cut lengthwise into two equal sections, which are operated by the two steel screws with thumb nuts as seen in the Cut.   These sections can be opened wide enough to let the Auger shank pass in without removing the nut on the end. This Handle is very neat and substantial and will be much in demand wherever introduced.

Price, per dozen, made of Ash Wood.........................$6 00

No. 1, 14 inches long.   No. 2, 17 inches long.   Both of one price.

GOODELL'S PATENT ADJUSTABLE HOLLOW AUGER.

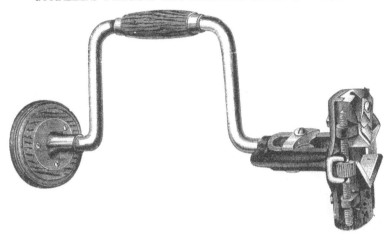

This auger is believed to be an improvement on anything hitherto in use. It has all the strength, durability and quick adjustability of any other, and it also has to drive it a Steel Bit Brace Sweep of the same size and finish as our No. 10 Brace, to wit: 14 inch sweep. As the Brace Sweep is fitted to the Auger it must always work entirely true, which is not the case when used in an ordinary Brace. Besides, it often happens that the Bit Brace on hand is not large enough to drive a spoke Auger. Adjustable to Cut from $\frac{1}{4}$ to $1\frac{1}{4}$ inches.

Price, each, with Sweep........ ... .\$4 00

SPOKE TRIMMER.

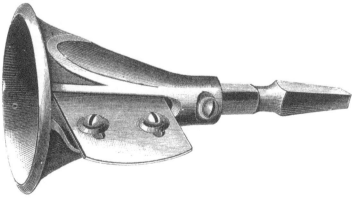

Price.................................... .. .. per dozen, \$6 00

FOOT POWER FAMILY GRINDSTONE.

After much experimenting we have now fully perfected our Grindstone for family use, and offer it to the public with a FULL GUARANTEE that it is a perfect machine; and also that it will please evey one who buys it. So far as we know it is the first Foot-Power Machine which has been fully adapted to the wants of families for household work, and of mechanics for grinding small tools. The stone is of the best quality and runs perfectly true. It is eight inches in diameter, one and a-half inches thick, and made at the Huron Quarries expressly for this use.

The EMERY WHEEL is the same size as the stone, and double coated on the side and rim with best Wellington Mills Emery. When not in use it is taken off and laid aside. A sponge is fastened in the side of the trough to keep the stone from throwing water when running at a high speed.

The Machine is run with a clutch, so that there can be no dead centres; but when the foot touches the treadle it starts off in the right direction, and runs at a very high or very low rate of speed, as desired. For grinding Carving Knives and all light tools, and for polishing Cutlery this machine is perfect. The legs are made to fold up for shipping, so as to occupy a small space. Weight, 26 pounds.

Price, including box...$3 50

JEWELERS' SAW.

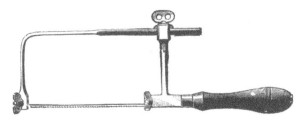

This Saw is made of forged spring steel, with rose-wood handle. It is nicely polished and nickel-plated, and will hold a saw from one to six inches in length.

Price..per dozen, $9 00

THE GEM SOLDERING CASKET, No. 1.

The above useful article is intended for family use, also for Farmers, Store-keepers, Mechanics, and every one using tin, brass, copper-ware, &c. The Casket contains Soldering Iron, Scraper, large bar of Solder, box of Soldering Salts and full directions, all enclosed in Fancy Box.

Received Centennial Medal, Diploma and Award of Merit.

The Gem Soldering Iron is so arranged as to fit any ordinary gas burner, and will heat in two or three minutes. Where there is no Gas the iron can be heated in the stove.

The gem Soldering Casket is without a rival or an equal, for simplicity, utility, economy, and convenience. Any housekeeper knows the annoyance of waiting for a tinsmith to repair an article. In mending two or three articles the Little Gem pays for itself.

The construction is so simple that a child can use it. In fact, it is a perfect marvel of usefulness, a durable compact, cheap, simple invention that will soon be found in the hands of every economical housekeeper.

Please do not confound this article with liquid solder, patch solder, &c., &c., as you cannot solder without a soldering iron.

No. 2.
SOLDERING IRON & SCRAPER

As seen in the cut. Put up in wooden boxes. Not adapted to gas burners.

BOX OF ROSIN.   BAR OF SOLDER.

Price, per dozen sets, No. 1.... $12 00
    "   "   "   2...... 6 00

GOODELL'S PATENT IRON LEVEL.

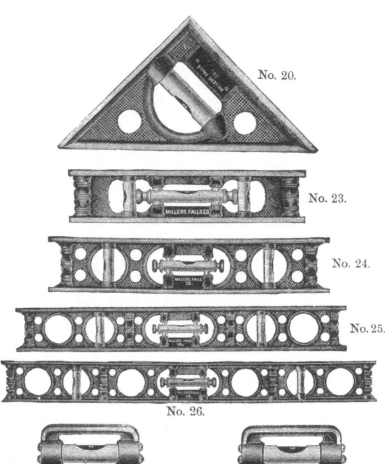

No. 20.

No. 23.

No. 24.

No. 25.

No. 26.

No. 21.   No. 22.

These levels are entirely new, both in design and construction. The improvement for which we shall ask Letters Patent consists in the ingenious method of adjusting the glasses, which is done by loosening one screw and tightening another. The points of these screws strike the case which holds the bubble glass near the end, one being on the upper, and the other on the lower side. The stocks are Japanned, while all the trimmings are Burnished Brass or Nickel-plated.

PRICES.

No. 20.— 4 inch Right Angle Triangle Level .... per doz.	$9 00
" 21.— 4 " Bench Level .... "	6 00
" 22.— 5 " " .... "	7 20
" 23.— 6 " Double Plumb Level .... "	18 00
" 24.—12 " " " " .... "	21 00
" 25.—18 " " " " .... "	24 00
" 26.—24 " " " " .... "	30 00

STRATTON'S PATENT LEVELS.

No. 1.

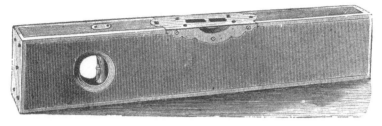

The engraving represents our No. 1 ADJUSTABLE PLUMB and LEVEL. The corners are protected from injury, and preserved true and perfect by Brass Rods extending the entire length of the Level, securely attached to the wood. These Levels have two ornamental Brass Lipped Side Views, heavy Brass Top and End Plates, and are made of the best, thoroughly seasoned Mahogany and Rosewood, Polished. Patented, March 1, 1870, and July 16, 1872.

PRICES.

	ROSEWOOD.	MAHOGANY.		ROSEWOOD.	MAHOGANY.
30 inch	$8 00	$5 40	24 inch	$7 00	$4 85
28 "	7 60	5 15	22 "	6 50	4 70
26 "	7 25	5 00			

No. 2.—Patent Adjustable Mahogany Plumb and Level, 2 ornamental Brass Lipped Side Views, heavy Circular End Top Plate, Tipped, Polished, 26, 28 and 30 inch.. 3 60

No. 3.—Is like No. 2, with Tips like No. 4; 26, 28 and 30 inch............ 3 00

No. 4.—Patent Adjustable Mahogany Plumb and Level, two Side Views, heavy Circular End Top Plate, Tipped, Polished, 26, 28 and 30 inch.............. 2 10

No. 5.—Patent Adjustable Cherry Plumb and Level, two Side Views, heavy Circular End Top Plate, Polished, 26, 28 and 30 inch.................... ... 1 50

No. 6.—Mason's Patent Adjustable Mahogany Plumb and Level, two ornamental Brass Lipped Side Views, heavy Circular End Top Plate, Polished, 36 in. 3 25

No. 7.—Mason's Level Mahogany Adjustable Level and Plumb, two side views, heavy Circular End Top Plate, Polished, 36 inch 2 50

No. 8.—Mason's Level Cherry Adjustable Level and Plumb, two side views, heavy Circular End Top Plate, polished, 36 inch..................... ... 2 00

No. 10.—Machinist's Level. The Machinists' Levels have Brass Side Views, heavy Brass Top and End Plates and Brass Corners. Rosewood polished. Patented March 1, 1870.

PRICES.

	ROSEWOOD.		ROSEWOOD.
12 inch	$2 60	8 inch	$2 00
10 "	2 40	6½ " without Plumb	1 10

No. 11.—Pocket Level. Brass, embossed and Nickel-plated. 6 inches long, ⅝ inch square ....each, 1 50

No. 12.—Pocket Level. Brass, Nickel-plated. Cylindrical, with base plate; round ends. Each, 2½ inch...... .... 0 75
" 3½ " 0 90

CHISEL GRINDER.

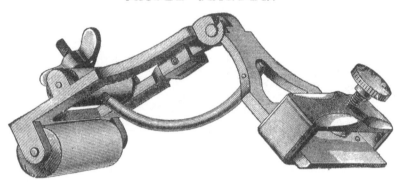

This is a new invention for holding Chisels, Plane Irons, &c., while grinding them. When put in the holder and brought to the right bevel with the adjusting screw, nothing is left to do but to bear it on the stone, and it will grind all right without further care.

Price.................................... per doz., $9 00

BIT GAUGE.

This cut shows the gauge in all of its parts. It will be seen that one bolt with thumb-screw tightens the clamps on the gauge spindle and auger bit at the same time. It will fit any size bit, and exactly gauge the depth of hole to be bored.

Price.................................... per doz., $3 00

GOODELL'S SPOKE SHAVE.

The circular shape of this tool enables it to work in smaller circles than other shaves.

The angle of the knife is such that it cuts instead of scraping the grain of the wood. Either handle can be removed to work in cramped places.

Price. .... per doz., $9 00

MONCE'S NOVELTY GLASS CUTTERS.

Patented July 1st, 1873.

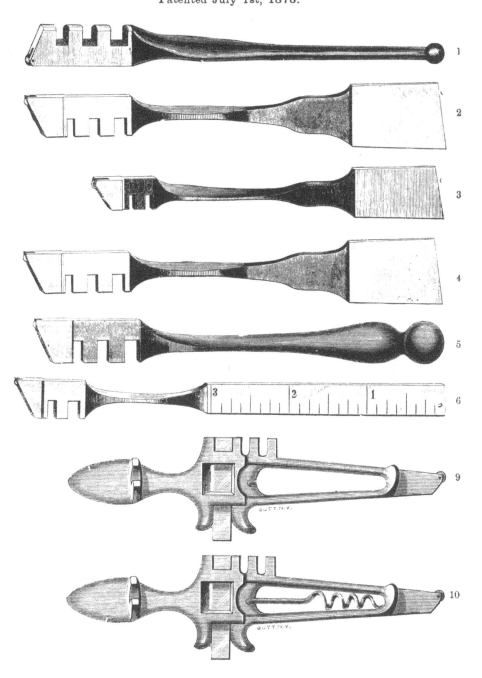

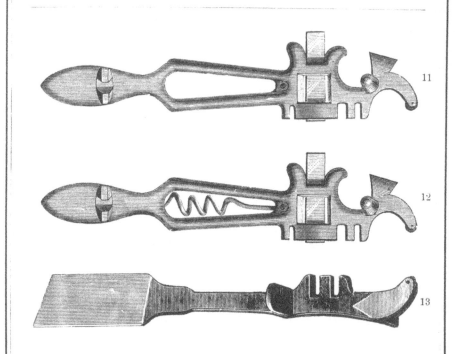

NEW STYLE GLASS CUTTER.

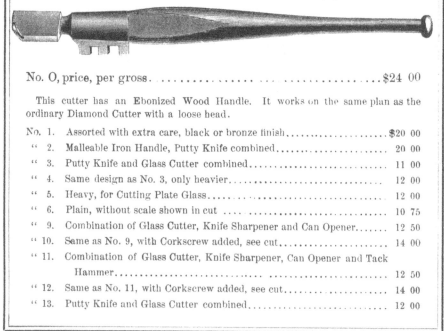

No. O, price, per gross $24 00

This cutter has an Ebonized Wood Handle. It works on the same plan as the ordinary Diamond Cutter with a loose head.

No. 1.	Assorted with extra care, black or bronze finish	$20 00
" 2.	Malleable Iron Handle, Putty Knife combined	20 00
" 3.	Putty Knife and Glass Cutter combined	11 00
" 4.	Same design as No. 3, only heavier	12 00
" 5.	Heavy, for Cutting Plate Glass	12 00
" 6.	Plain, without scale shown in cut	10 75
" 9.	Combination of Glass Cutter, Knife Sharpener and Can Opener	12 50
" 10.	Same as No. 9, with Corkscrew added, see cut	14 00
" 11.	Combination of Glass Cutter, Knife Sharpener, Can Opener and Tack Hammer	12 50
" 12.	Same as No. 11, with Corkscrew added, see cut	14 00
" 13.	Putty Knife and Glass Cutter combined	12 00

Improved Patent Universal Angular and Ratchet Drilling Machine.

We have for sale Morse Twist Drills from ¼ to ¾ inch, with uniform ½ inch shanks to fit this machine.

They will work at any angle. By placing the crank on the drill spindle, it will work with a ratchet or without. We send a chuck with each machine, which will hold 1-16 to ½ inch drills.

PRICE LIST.

No. 1. Weight, 26 lbs. Drills up to ⅝ inch hole.........$20 00
" 2. " 52 " " " ¾ " " 25 00
" 3. " 106 " " " 1½ " " 40 00

The No. 2 Drill has two sets of gears, making either speeded or geared back machine.

These Drilling Machines are now made of steel, and are first-class in all respects. For repair work in mills they are almost indispensable, as they can be attached to a broken machine without taking it apart, and swung round to drill at any angle.

TWIST DRILLS.

We have constantly in stock Twist Drills (see page 11), which we will send by *mail, postage paid,* upon receipt of price, or they can be bought from dealers selling our Vise and Anvil Drill and Drilling machines.

CARVING TOOLS.

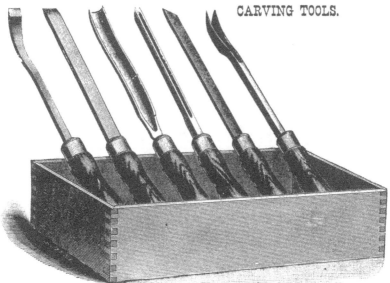

Six Tools in a Wooden Box. Price, per dozen sets....$12 00

These Tools are forged from best quality of steel, and are fully warranted. They have Rose-wood Handles and are sharpened ready for use.

JOHNSON'S PATENT BENCH HOOK.

No. 2.

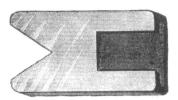

The piece marked 2 slips on to the stop and is used to hold boards when standing on edge to be planed, and other like purposes.

This hook is of the most substantial character, being positive in its action. It fits any thickness of bench, and is fitted in by boring a 1½-inch hole. The stop is made of forged steel, and can be turned so as to present either of its four faces. The shaft is of wrought iron, and screwed into the stop. All parts are turned in a lathe, and those parts which come to view are polished. It is the most substantial, and most easily fitted bench hook which we have yet seen.

Price. .... per doz., $12 00

CRISPIN'S AWL.

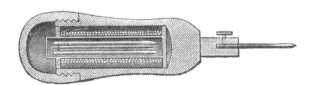

This is a tool for everybody. It is a hollow handle made of hard wood, five inches long, and weighing three ounces. It contains inside, on a spool, fifty feet of best waxed linen shoe thread. The spool is also hollow, and contains three awls and five needles of various shapes and sizes. The thread fits the needles, and the awls fit the handle, and are held by a set screw, as seen in the cut. It is for use in the house, stable, field, camp, or on the road, for making immediate repairs, when one use of it will be worth more than its whole cost.

Price. .... per doz., $3 00

MILLERS FALLS BORING MACHINE.

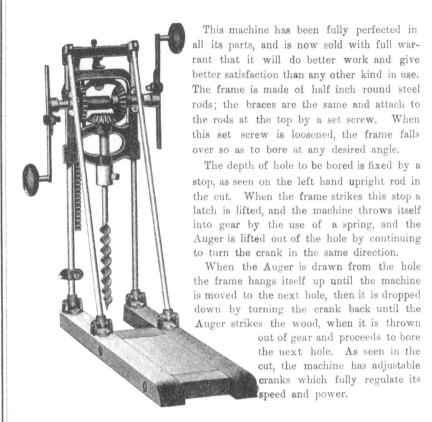

This machine has been fully perfected in all its parts, and is now sold with full warrant that it will do better work and give better satisfaction than any other kind in use. The frame is made of half inch round steel rods; the braces are the same and attach to the rods at the top by a set screw. When this set screw is loosened, the frame falls over so as to bore at any desired angle.

The depth of hole to be bored is fixed by a stop, as seen on the left hand upright rod in the cut. When the frame strikes this stop a latch is lifted, and the machine throws itself into gear by the use of a spring, and the Auger is lifted out of the hole by continuing to turn the crank in the same direction. When the Auger is drawn from the hole the frame hangs itself up until the machine is moved to the next hole, then it is dropped down by turning the crank back until the Auger strikes the wood, when it is thrown out of gear and proceeds to bore the next hole. As seen in the cut, the machine has adjustable cranks which fully regulate its speed and power.

The gears are all cut, which is not common in other machines.

We do not desire the reputation of making the cheapest goods in market, but we mean to deserve the name of making the best.

PRICES.

Machine, without Augers..... ....$7 50

Augers in sets............18.............23............41 quarters.

$4 50 $5 50 $10 25

Sizes of Augers..(1, 1½, 2.)..(1, 1⅓, 1½, 2.)..(½, ⅝, ¾, ⅞, 1, 1¼, 1½, 1¾, 2.)

DOUBLE CRANK BREAST DRILL.

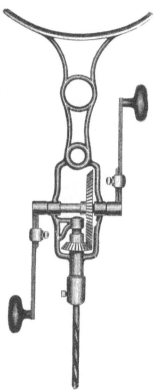

To meet the demand for a more powerful and more durable breast drill, we now offer this Double Crank Machine, No. 14. It has very heavy cut gears, speeded about two to one, and double cranks, which are adjustable to any length, and by which the power and speed may be regulated as desired.

The Breast Plate is twelve inches long, and curved to fit the shape of the body, thus preventing it from swinging around when in use. The hole in the spindle which receives the Drill is one-half inch in diameter. We sell Morse Twist Drills of all sizes with one half-inch shank to fit this machine. We also make a chuck to fit it which will hold all round and square drills below 7.16 inch. This is a tool which will stand very hard usage, and never wear out.

PRICES.

Double Crank Breast Drill, No. 14.............................$5 00
Adjustable Chuck to fit same.................................. 1 00
      Morse Twist Drills to fit this machine (See page 11).

HAND FORGED ROUND SCREW-DRIVER BIT.

There is a special demand for a Screw-driver Bit which will stand a great strain. Nothing hitherto in the market would fill the bill. We have made these drivers for special customers during the last two years, but hesitated to put them in our Catalogue on account of their high price. But many mechanics are willing to pay the price, and so we offer them at the following rates:

No. 1, 5 inches long............. ....per dozen, $1 50
  " 2, 8 "   "   .... "   2 00

"GREENFIELD" FORGED OX SHOE.

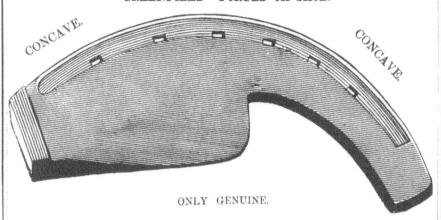

CONCAVE.   CONCAVE.

ONLY GENUINE.

Made under the Parker and Colburn Patents, from Burden's H. B. and S. Iron. Nail holes punched and every shoe perfect.

The Parker and Colburn Patents cover broadly the dies in which the shoes are forged. We are the *only* licensees, and all parties are cautioned against using either of the dies or the forging mechanism or processes so protected, as our rights under said patents will be fully maintained.

While we can furnish either the *Concave Shoe* with *One Calk*, or the *Flat Shoe* with *Two Calks*, we emphatically recommend the Concave, with one Calk, for the following reasons. viz.:

First—Because the entire bearing of the shoe should come upon the *shell* of the hoof and not upon the *ball* or tender part of the foot, as is necessarily the case with the flat shoe. This principle is recognized by all experts in the shoeing both of oxen and horses, and will prevent a tendency to sore footedness.

Second—Because by having one calk only the shoe can be cut off or lengthened and fitted more perfectly to the foot.

Third—Because by having one calk only the shoer *can make the other calk at any angle he desires.*

No. 0, Full Length, Concave, 4½ inches, weight per set of eight shoes, 2 pounds.
  " 1,   "   "   "   5   "   "   "   "   3   "
  " 2,   "   "   "   5½   "   "   ".   "   3½   "
  " 3,   "   "   "   6   "   "   "   "   4   "
  " 4,   "   "   "   6½   "   "   "   "   5   "

Packed in boxes or kegs of 100 pounds, half each rights and lefts. Full weight, and no charge for packages.

PRICES.

For orders of One Ton, or more............................ ...... 9 cents per pound.
    "   1,000 lbs. or more............................ 9½   "   "
    "   500   "   " 10   "   "
    "   Less than 500 lbs............................10½   "   "

Terms, Net Cash, 30 days.

Made ONLY by MILLERS FALLS COMPANY.

Formerly Forged by Greenfield Tool Company.

TOOL HOLDER.

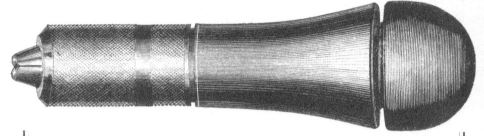

This cut represents the general appearance of all the four kinds of Tool Handles which we make. They are made of Rose-wood or Cocobola (except No. 2) with Lignumvitæ Caps, highly polished and of beautiful appearance. The ferrule and jaw are heavily Nickel-plated.

The jaws will hold, not only the tools contained in the hollow handle, but all other things, from a needle to a mill-file. No other Tool Handle in market will do this. It also answers the purpose of a hand vise. The handles are much larger than this cut, the length being as follows, No. 1, 6, No. 2, 5½, No. 4, 6½, No. 5, 7½ inches.

The Tools are made from Steel of the highest grade, tempered by men of great experience, honed to a fine cutting edge, and are all highly finished. They are made for service and will give the greatest satisfaction. The Jaws in the handle shut over the shoulders of the tools (as seen in the cuts) so as to make it impossible to pull them out when in use.

The Tools in each holder are about the size represented in the cuts. These goods bear the highest price of any in market, but their quality is as much higher as the price.

No. 1, per dozen............. ....$12 00
" 2, Solid Maple Handle, per dozen........................ 4 00
" 4,   " 12 00
" 5,   " 18 00

TOOLS FOR No. 1 HANDLE.

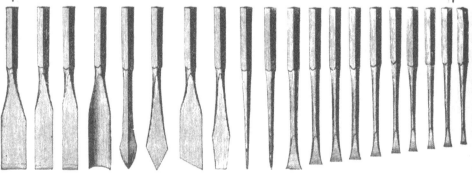

TOOLS FOR No. 4 HOLDER.

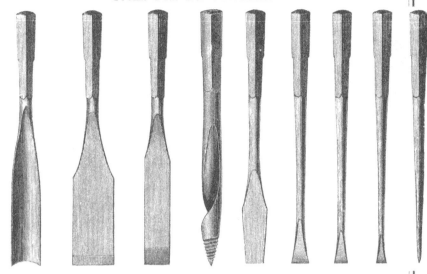

TOOLS FOR No. 5 HOLDER.

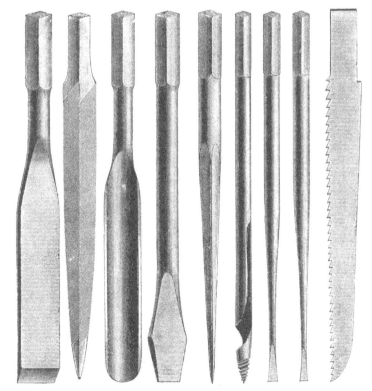

STAR ★ SAWS.

It is cheaper to buy a new saw than to file an old one. As our saws are never to be filed they are tempered so high that one will last four times as long as an ordinary saw, with one filing. This will be a very great saving in money, and also in the delay of filing. Look at the prices below and you will never file another saw of these kinds.

We fully warrant all saws marked with a star and bearing our name.

These saw blades are all as they came from the tempering furnace. Their strength and temper would be much injured by any subsequent polishing.

Take notice that Blades and Frames are listed separately, and order what you want of each kind.

STAR HACK ★ SAW BLADES.

Length of Blade, 6, 7, 8, 9, 10, 11 12; assorted, 6 to 9.
Price, per dozen, 55c. 60c. 65c. 70c. 85c. 95c. $1.05   65c.

STAR HACK ★ SAW FRAMES.

No. 0 extension frame, to hold 10, 11 and 12 inch, steel polished
and nickeled..per dozen, $12 00
No. 1 extension frame, to hold 6, 7, 8 and 9 inch, steel polished
and nickeled..per dozen, 9 60

No. 2 solid frame, to hold 8 inch, steel polished and nickel-
plated...per doz., $8 40

This No. 4 is a Patent Cast-iron Frame carrying a 9 inch blade, and so constructed as to face it in four different directions. The pins which hold the blade are fast in the frame and cannot drop out.

It is a very stiff and desirable Frame with Japan finish.

Per dozen...$4 80

This is our latest improved Patent Steel Frame, the heaviest and stiffest which we make. It is highly polished, heavily nickel-plated, and will face the blades in four directions.

It also has our patent pins which cannot drop out. The handle is ebonized.

No. 5 will hold blades 6, 7, 8 and 9 inches long....per doz., $9 60
  "   6   "   "   "   9, 10, 11 " 12   "   " .... "   12 00

This is a Wooden Frame, made of second growth Ash. It is strong and substantial, carrying a 9 inch blade, and facing it only one way. It will do good work and costs but little.

Price..per dozen, $2 00

KEY HOLE SAW.

These Patent Key-hole Saw blades are 8½ inches full length. Like the Compass saws they will cut much faster than any other kind.

The handles are made of hard Maple, turned and polished. They are so constructed that the blade can be pushed back inside, so that when cutting a small circle only the point will project.

Price of handle..per dozen, $1 25
      "   blades.......................................   "   70

STAR ★ BUTCHER SAW.

STAR BUTCHER SAW.

Flat crucible steel backs, highly polished and heavily nickel-plated.   Beech handle, polished edges, three brass screws.   This is the best butcher's saw frame in any market.

Inches,   14,   16,   18,   20,   22,   24.
Per dozen...$19 00, $20 00, $21 00, $22 00, $23 00, $24 00.

STAR BUTCHER ★ SAW BLADES.

Length.	Width.	Gauge.	Teeth to inch.	Per dozen.
14 and 16 inches.	$\frac{1}{2}$ inch.	24	$9\frac{1}{2}$	$1 08
18 " 20 "	$\frac{5}{8}$ "	24	$9\frac{1}{2}$	1 20
22 " 24 "	$\frac{3}{4}$ "	24	$9\frac{1}{2}$	1 32

STAR KITCHEN SAW.

STAR KITCHEN SAW.

These Steel Frames are polished and nickel-plated, with Beech Handles.   They have what most other kitchen saws do not have, an arrangement for straining the saw, which adds much to their value.

Price of 14 inch frames, with one blade......per doz., $6 00
       "   14   "   blades................ ...   "   1 00

COMPASS SAW, No. 1.

This Compass Saw has a Beech Handle which is so constructed as to face the blade in four different directions.   By loosening the thumb nut the blades are easily removed.   The blades are patented, and will cut very much faster than any other kind in market, especially with the grain of the wood.

Price of handle, without blade ....per dozen, $3 00

COMPASS SAW, No. 2.

As seen in the cut, this No. 2 handle is so made that the blade can be pushed back through it, so that only the point shall project when small circles are to be cut.

Price of No. 2 Handle...............................per dozen, $2 50

Compass saw blades are put up six in a package, two each, 8, 10 and 12 inches long, not including the shanks which go in the handle.

Price of six blades.............................. ....50 cents.

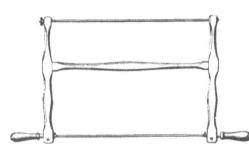

TURNING WEB.

This frame is made of birch wood, with ebonized handles. There is an index on each handle to show the operator just how far to turn each. The friction is regulated by screws. It is quite superior to most other kinds in market.

Frame, with one blade, each, $1 00

18 inch blades...per dozen, 2 00

PRATT'S BLIND OPENER.

This is a simple device for opening and closing Blinds and Shutters, without opening the window. In common with many others, we have long experimented with these Blind Openers, but in advance of any others we happened to strike the right thing, and the only thing thus far, which will do the work to everybody's satisfaction.

Whatever hinges are on the blinds, they are not to be disturbed. A screw eye or staple is placed in the middle rail of the blind, and a lever working in this staple moves the blind either way. This opener locks the blind fast at any point where it may be when you let go of the crank.

They are put on without going outside of the room and in a very few moments' time. Full directions go with each set.

The crank and face plate, which remain inside the room, are highly ornamental, being polished and nickel-plated

Price, per dozen set..$9 00

LANGDON ADJUSTABLE MITRE BOXES.

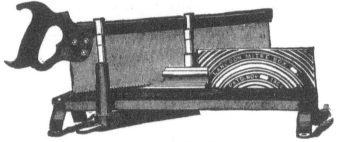

Sizes A, 1, 2, 2½ and 3.

Order by the No.

No. 1 is size A, with No. 4 Saw, for wood, 15 x 2½	....	$ 6 50
" 2 " " " metal, 14 x 2½	....	6 50
" 3 " " " with both Saws	....	8 00
" 4 is size 1, with No. 6 Saw, for wood or metal 18 x 4	....	9 00
" 5 " " " " 20 x 4	....	9 50
" 6 " " " " 22 x 4	....	10 00
" 7 " " " for wood, 24 x 4	....	10 50
" 8 " " " " 26 x 4	....	11 00
" 9 is size 2, " " " 22 x 4	....	11 00
" 10 " " " " 24 x 4	....	11 50
" 11 " " " " 26 x 4	....	12 00
" 12 is size 2½, " " " 28 x 5	....	14 00
" 13 is size 3 " " " 24 x 6	....	18 50
" 14 " " " " 26 x 6	....	19 00
" 15 " " " " 28 x 6	....	19 50
" 16 " " " " 30 x 6	.. ..	20 00

We are occasionally asked to sell Mitre Boxes *without Saw.* The following Mitre Boxes will be sold *without Saw*, if desired, but without guarantee, as Mitre Boxes cannot be depended upon to work with saws indiscriminately furnished, and we cannot afford to sacrifice our reputation, which has cost us years of labor to establish.

	No. 1.	No. 2.	No. 2½.
New Langdon.....................$	7 00	8 00	10 25
Langdon and Improved............	8 25	9 25	11 25

*No. 3 $16 25. No. A, $5 25.

Parties ordering Mitre Boxes who have Saws which they wish to save should send them to us to be adjusted to the Box in which they are to be used. No charge will be made for the adjustment to the Box in which they are to be used, unless money is paid for *putting the Saw* in proper working order.

Such adjusted boxes will be guaranteed to do good work if the Saw is found good. Prices subject to change without notice.

Size A gives 3½ inches width at right angles, 2¼ at mitre. Weight, 5½ lbs.
" 1 " 6 " " " " 4 " " 10½ "
" 2 " 9½ " " " " 6½ " " 11½ "
" 2½ " 9½ " " " " 6½ " " 12½ "
" 3 " 9½ " " " " 6½ " " 25 "

NUMBER OF PARTS OF LANGDON MITRE BOX.
MADE TO INTERCHANGE.

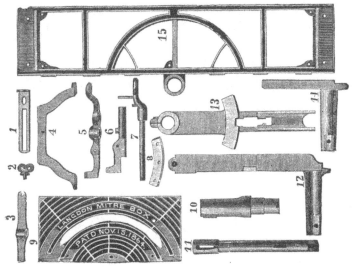

☞ In ordering Backs (No. 9) please state whether a Right or Left-hand back is wanted.

List of Parts (as shown in cut) and Price of same.

All parts of the Mitre Box are made to interchange, and will be sent by mail, if postage accompanies order, otherwise by express. Postage, one cent for each ounce.

Persons ordering parts of Machines will please state the number of the piece wanted, and the size of the machine for which it is designed, and enclose the price and postage for the same.

		Mitre B Size 1.	Mitre B Size 2.	Mitre B Size 2½	Mitre B Size 3.
1 Stop Guage	each	$0 10	$0 10	$0 15	$ 020
2 Thumb Screw	"	1 0	10	10	10
3 Back Lever	"	15	15	15	30
4 Legs	"	15	15	15	30
5 Stop Lever	"	30	. .	. .	60
6 Stop Lever	"	. .	45	45	. .
7 Thumb Lever	"	. .	20	20	. .
8 Gib	"	25	25	25	50
9 Back	"	30	30	50	60
10 Back Posts	"	35	35	60	70
11 Saw Guides	"	55	55	90	1 10
12 Swinging Lever	"	90	. .	. .	1 80
13 Swinging Lever	"	. .	1 00	1 00	. .
14 Sliding Post	"	. .	50	85	. .
15 Bed	"	1 00	1 00	10 0	20 0
Bottom Board	"	10	10	5	20
Spiral Spring	"	5	5	5	5
Screws	"	5	5	5	5
Lever Pins	"	5	5	5	5
Set Screw	"	. .	10	10	. .
Clamp	"	. .	25	5	. .

THE NEW LANGDON MITRE BOX.

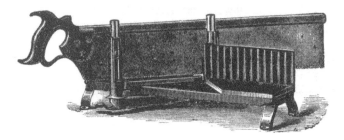

ARRANGED FOR BACK SAW.

We warrant all our Mitre Boxes to do perfect work when supplied with Back Saws fitted by us.

Order by the No.

Sizes B, 1, 2, and 2½.

No. 16½ is size B, with Saw,			12 x 2........ .. $4 50
" 17 " 1, " No. 6 Saw, for wood,			18 x 4........... 7 75
" 18 "	"	"	20 x 4........... 8 25
" 19 "	"	"	22 x 4........... 8 75
" 20 "	"	"	24 x 4........... 9 25
" 21 "	"	"	26 x 4........... 9 75
" 22 is size 2,	"	"	22 x 4........... 9 75
" 23 "	"	"	24 x 4...........10 25
" 24 "	"	"	26 x 4...........10 75
" 25 is size 2½	"	"	28 x 5...........12 00

For prices of Boxes without Saws, see page 52.

DIRECTIONS.

The positive angles of the NEW LANGDON, as notched in the segment, are the same as in the Langdon Mitre Box.

The NEW LANGDON has the *additional* feature of easy adjustment to *any possible angle desired.*

Find the angle wanted by setting a bevel as required, then move the saw so that the angle formed by the saw and the back of the box shall conform to the bevel, fasten the saw in place by turning down the screw in the grooved rest or gib, directly above the letter A, as shown in the cut. This will hold the saw strongly in place for the time being, and a turn of the screw backward will place it in order again for the segment angles.

PARTS AND PIECES OF NEW LANGDON MITRE BOX.

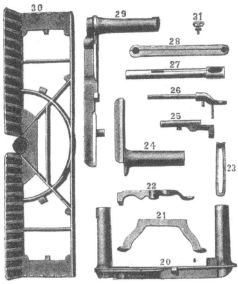

Persons ordering parts will please state the number of the piece wanted, and the size of the machine for which it is designed, stating whether it is for the Langdon or New Langdon.

	No. 1.	No. 2.	No. 2½.
20 Swinging Lever.............................	$1 25	$ ..	$ ..
21 Legs.....................................	15	15	15
22 Stop Lever...............................	30	..	..
23 Stop Gauge..............................	10	10	10
24 Sliding Post.............................	..	50	85
25 Stop Lever..............................	..	45	45
26 Thumb Lever............................	..	20	20
27 Guide...................................	55	55	90
28 Gib.....................................	25	25	25
29 Swinging Lever.........................	..	1 25	1 35
30 Bed and Back	1 45	1 45	1 60
31 Thumb Screw...........................	10	10	10

THE ALEXANDER ADJUSTABLE JOINTER GAUGE.

We claim for this Tool:

1st.—The ease with which it is attached to any Plane, Jack or Jointer.

2d.—The great amount of time saved, without the use of Bevel or Try Square.

3d.—Is adjustable for squaring, or to any Bevel desired.

4th.—The work completed in the highest state of perfection, even by inexperienced workmen.

Patented July 16, 1872.

Attach the Guage to the plane so that the depression in the centre of the wing of the Guage shall be directly opposite the cutting edge of the plane iron.

Price, per dozen...$18 00

NEW LANGDON MITRE BOX IMPROVED.

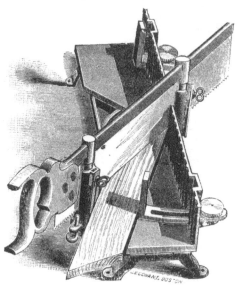

Ordinary Mitre Boxes, cut from right angles to 45 degrees inclusive, as shown on the diagram, A, B, C.

The New Langdon Improved cuts, by using the circular arms or guides, from right angles to 73 degrees on 2½ inch stuff, as shown on diagram A, B, D, varying more or less with width of stuff.

The only box adjustable for mitering circular work in patterns, emery wheels and segments of various kinds.

Sizes 1, 2 and 2½.

Order by the No.

No. 26 is Size 1, with No. 6 Saw, for wood, 18 x 4 . . . . $ 9 00
   "   27   "   "   "   20 x 4 . . . .   9 50
   "   28   "   "   "   22 x 4 . . . . 10 00
   "   29   "   "   "   24 x 4 . . . . 10 50
   "   30   ."   "   "   26 x 4 . . . . 11 00
   "   31 is Size 2,   "   "   22 x 4 . . . . 11 00
   "   32   "   "   "   24 x 4 .. . . 11 50
   "   33   "   "   "   26 x 4 . . . . 12 00
   "   34 is Size 2½,   "   "   28 x 5 . . . . 14 00

For prices of these Boxes without Saws, see page 52.

THE ROGERS MITRE PLANER.

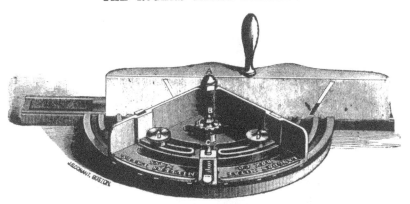

Patented September 19th, 1882.

It is made entirely of iron, and is arranged for planing any desired angle on straight or curved work.

The Main Bed Piece is semi-circular in form with a way or frame at its rear, on which the plane runs.

The upper, or movable Bed Plate is in Quadrant form, having at right angles sides which act as guides for the material to be planed, and revolving on a pivot A, at the end, enables the user to form the desired angle for straight work, and place it in its proper position against the face of the plane. When the Quadrant or movable Bed Plate is in the center of the main bed piece, its side elevations form an exact mitre, so that no change is required in planing the ends of parts for frames of four sides.

In the sides of the Quadrant are two adjustable guides or rests, kept in position by set screws, D. D.

The special object of these rests is to enable one to finish the ends or angles on curved work with exactness.

In preparing pieces for circular or oval work, frames, pulleys, emery wheels, circular patterns, etc., it is necessary to plane the ends of the various segments at varying angles. In planing these the point of the quadrant near the plane and the adjustable guides form the rests required for accurate work.

The Quadrant is kept in position at any angle desired by pressing the catch, C, down into the notches prepared for it, or by the thumb-screw, B, and can be used in connection with the arms or guides, as desired.

The Planer is manufactered in three sizes, each supplied with iron planes.

Size 2, carrying plane irons, 2 inches wide, sold for.. $20 00
" 3½,   "   3½   "   "   .. 25 00
" 4,   "   4   "   "   .. 30 00

These goods are warranted.

POWER'S WAGON JACK.   PELICAN NAIL PULLER.

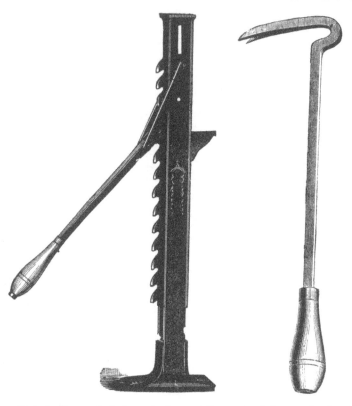

Power's Wagon Jack is made to meet the wants of the regular hardware trade. It is very compact, and occupies but little space in transportation, or in the store. It is made of wrought and malleable iron. When the lever is depressed it is carried past the center, so as to be self-locking. The lift is from 12 to 32 inches. It is the quickest working and most durable jack in market, and also the handsomest, being nicely Japanned and ornamented.

	Lift.	Length of Lever.	Will raise.	Weight.	Price.
No. 1—	12 to 32 in.	13 in.	1,000 lbs.	11 lbs.	$18 00 per doz.
" 2—	14 to 34 in.	16 in.	1,500 lbs.	16 lbs.	27 00 "

The Pelican Nail Puller is the least expensive, yet the most rapid working nail puller in use. It is made of the best tempered steel and is warranted to do perfect work. It weighs only one pound and takes but little room in a carpenters' chest. To be worked with a hammer. We regard it as one of the most useful tools which we have recently put upon the market.

Price.. per dozen, $9 00

SOLID STEEL ANVILS.

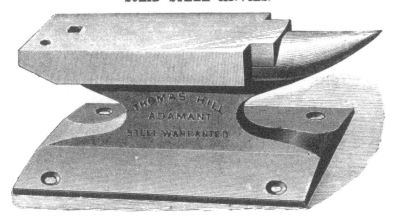

These Anvils are made in this country, and tempered by Peter Wright's old Foreman. They are warranted to be perfect in all respects.

No.	Long.	Wide.	Weight.	Price.
1	4½ inches.	1½ inches.	4 lbs.	$1 60
2	5 "	1½ "	5 "	2 00
3	6 "	⸺ "	7 "	2 50
4	7 "	2 "	9 "	2 75
5	8 "	2 "	10 "	3 00

The Horn is 2 inches long in addition to the length of face given above. The face, sides and horn are nicely polished. The bottom part is painted red.

HAM STRINGERS.

These Ham Stringers are made of fine tempered steel, polished and nickel-plated. Entire length, 12 inches.

Price, per dozen $4 00

DUPLEX SCREW DRIVER, No. 2.

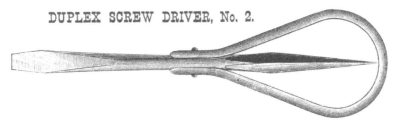

Screw-Driver and Reamer combined. Both Bit and Handle made of tempered Steel, polished and Nickle-Plated.

Price, per dozen.................... .... $3 00

No. 1 RATCHET SCREW-DRIVER.

This is a Double Pall Ratchet Screw-driver. The Handle is seven inches long, made of rose-wood or cocobola highly polished, and the iron part is heavily nickel-plated. With each handle is a set of three Bits, 5, 7 and 9 inches long, as seen in the cut. These are also polished and nickel-plated. It is in every respect as fine a tool as we can make.

Price, with Box..$21 00 per dozen.
Extra Bits for same....................................... 20 each.

No. 2 RATCHET SCREW-DRIVER.

This is a Single Pall Ratchet Screw-driver. Like the No. 1, it drives a screw in or out with the ratchet, or may be made stationary. The whole length is eleven inches. The Handle is cocobola, both bit and handle highly polished, and the bit nickel-plated as well as the metal part of the handle.

Price, with one Bit .... $12 00 per dozen.

DUPLEX SCREW-DRIVER, No. 1.

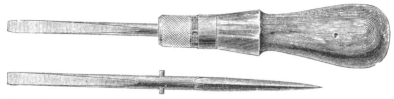

Generally when a Screw-driver is used, some tool is wanted to make a hole for the screw. This Square Reamer on the reverse end of the driver is better for that purpose than a gimlet or any other tool. The tool can be turned around instantly so as to use either end. It is hand-forged, carefully tempered, polished and nickel-plated. The handle is cocobola polished, with a nickel-plated chuck. This is entirely new and we judge will be a popular tool.

Price, complete...$6 00 per dozen.
   "   for Bits alone to fit Handle....................... 1 80   "

GOODELL'S RATCHET DRILL.

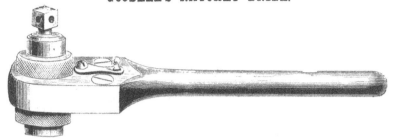

This is the latest Improved Ratchet Drill and the best finished one in market, being polished and nickel-plated throughout.

No. 1.	2.	3.
Lever, 10 in.	12 in.	14 in.
Price, each, $4 00	$6 00	$8 00

ADJUSTABLE WAGON WRENCH.

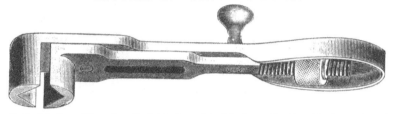

Malleable Iron Wrench, adjustable from $\frac{3}{4}$ to 2 inches. By the aid of a spring the wrench takes hold of the nut with a tight grip, so that it cannot be shaken out. As seen in the cut, there is a knob on the back side which is used as a crank when the wrench is made fast to the nut.

Price, per dozen.................. ....$3 00

THE ADAMS COMBINATION TOOL.

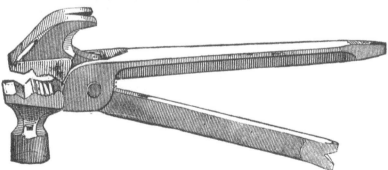

This Tool consists of a Claw Hammer, Gas Plyers, Wrench, Screw Driver, Tack Puller and Box Opener. Best grade of drop forged steel. Working parts polished. Warranted to be first class in all respects. Made in two sizes, as follows:

No.	Length.	Hammer Face.	Weight.	Price, per doz.
1	10 inches.	$1\frac{5}{16}$ inch.	23 ounces.	$12 00
2	9 "	$\frac{6}{16}$ "	14 "	9 00

GUNN'S TOOL CABINET.

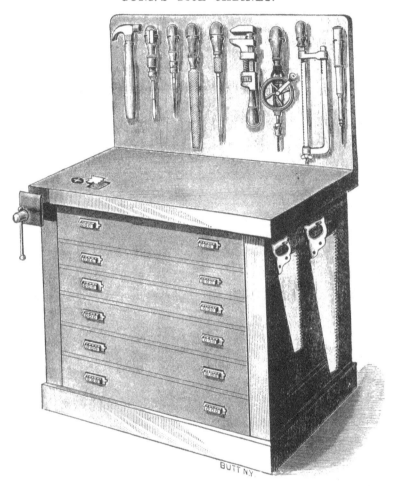

BUTT N.Y.

This is a Work Bench and Tool Chest combined, and is made for the use of those who have no room in the house for an ordinary work bench.   The Cabinet is 18 x 36 inches.   The top, or bench, projects one inch on the front and ends, being 19 x 38 and 34 inches high.   It is made of cherry or hard maple plank, two inches in thickness, and has a detachable Back Board 16 inches high with racks for tools as shown in the cut, but the tools are not included in the price, except the Vise and Bench Stop.

There are Six Drawers for holding Tools, Nails, Screws, Brads, etc., being much more convenient than any tool chest now in use.   The Vise has steel-faced jaws and is of a new pattern made expressly for this Cabinet,

We shall also make a finer Cabinet of the same size, with a cover like a sewing-machine.   This will be no more blemish to the best room in the house than a nice bureau.

Price of No. 1 Tool Cabinet (like cut), each............ ....$16 00
" 2 " with Cover, Ash or Walnut..............24 00

In due course, we expect to make a full line of Amateur Work Benches.

New York, May 1, 1887.

Not having room enough for our growing business at No. 74 Chambers Street, we have taken a full store with basements at No. 93 Reade Street, two blocks away. With a larger store and larger stock, we hope to serve our customers more promptly and with a greater variety of goods. As will be seen in this Catalogue, we have added a number of new tools to our list and have made many valuable improvements on the goods heretofore made by us.

We have not room in this Catalogue to even name the large list of Designs for Bracket Sawing which we publish, but will send miniature sheets of them all on demand. We are sole publishers of Morton's, Jennings' and Russell's Designs, 550 in number, and New York Agents for Hope & Ware's, 300 in number, and Pomeroy's, 50 in number.

We also furnish Adams & Bishop's Designs on demand.

We have manufactured most of the Hand and Foot Power Bracket Saws which have been used the world over during the past twelve years. With improved machines the trade is again rapidly increasing, and the goods are becoming staple in most hardware stores. We keep a full stock of Saw Blades of our own make, Wood, Designs and all things in the Bracket Sawing line.

In all departments we shall offer goods of the best quality and at reasonable prices.

MILLERS FALLS CO.,

No. 93 READE ST., NEW YORK.

INDEX.

No. 4 TOOL HOLDER.

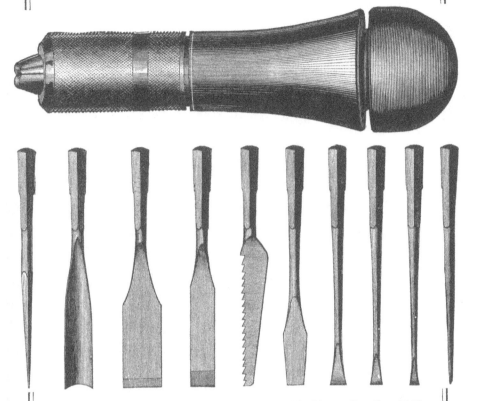

These Tool Handles are made of Rosewood with Lignumvitæ Cap, highly polished and of beautiful appearance. The iron ferrule and jaws are heavily Nickel-plated.

The Steel Jaws will hold perfectly, not only the Tools contained in the hollow handle, but all other things from a Needle to a Mill File. No other Tool Handle in the market will do this; it answers the purpose of a small Hand Vise.

These Cuts are about two-thirds the size of the Handles and Tools which they represent. The Tools are made from Steel of the highest grade, tempered by men of great experience, honed to a fine cutting edge, and are all highly finished. They are made for service and will give the greatest satisfaction. The Jaws in the handle shut over the shoulders of the tools (as seen in the cuts), so as to make it impossible to pull them out when in use.

No. 4 Handle and 10 Tools.................each, $1 00

MILLERS FALLS COMPANY,

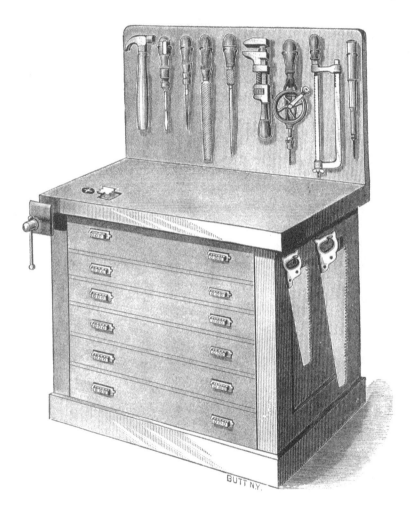

93 READE STREET,

NEW YORK.

Lightning Source UK Ltd.
Milton Keynes UK
UKHW041531081222
413606UK00020B/125